GW01451420

HARRIERS:
JOURNEYS AROUND
THE WORLD

A LIMITED EDITION OF 700

HARRIERS:

JOURNEYS AROUND
THE WORLD

A PERSONAL QUEST

For WALBY.

*HOPE YOU ENJOY THIS BOOK.
GOOD READING & BEST WISHES.*

Don Scott

*Don Scott.
8/11/08.
3/700.*

Tiercel
PUBLISHING

ISBN 978-0-9532002-6-9

Published November 2008
A Limited Edition of 700

Tiercel PUBLISHING 2 Mill Walk Wheathampstead Herts AL4 8DT

Contents

DEDICATION

For Linda and Douglas, and also my late parents William and Margaret

To the Harriers themselves – who were my inspiration for this book.

also dedicated to the memory of

Roger Clarke
1952–2007

Roger was a good friend whom I first met on 20th July 1991 through our mutual interest and passion for harriers. We shared many memorable 'Harrier' moments not only in my native Northern Ireland, but also during our visits to the Isle of Man, the UK, Holland and three 'never to be forgotten' trips to India – all in the name of harriers.

His unceasing writings culminated in three superb books on his favourite species and numerous papers, several of which I was honoured to have co-written with him. He was also a master tactician in the art of dissecting raptor pellets of many different species. Right up to his death, Roger was still working on another two important books, one an update of the Hen Harrier by Donald Watson, in which we jointly wrote a new chapter about our 17 years of study in Northern Ireland and the other, about pellet identification of diurnal raptors. Thankfully, these are to be posthumously published in the near future.

Due to the high esteem that I had for Roger, I went to visit him on 18th January 2007 shortly before his death and then attended his funeral service in his home village of Reach near Cambridge, on 7th February. His premature death, aged only 54, was not only a sad day for his wife Janis, son Mostyn and daughter Bethan, but for me also, and his many harrier colleagues at home and abroad. I regret that Roger did not live long enough to read my contribution, but he did manage to read a revised draft that he requested I bring with me when we sadly met for the last time in January. An email from him only hours before he passed away said – "I would not do too much to the text, after all, it's your own statement/style on life." I was delighted that the contents had not only met with his approval, but also his meticulous high standards.

I am not sure what it is about harriers that attract eccentrics, but there have been many who qualify. I think first of the late Eddie Balfour, who tramped the Scottish and Orkney moors after he fell under the spell of Hen Harriers. He got so carried away with them and their polygynous mating system that he ended up spending 40 years of his life researching them. In continental Europe, Erkki Korpimaki and before him Yngvar Hagen, monitored hundreds of raptors, including Hen Harriers, in Finland and Norway and found ever stronger harrier-vole cycles the further north they went. Fascinating stuff and now a part of the mainstream literature on predator-prey cycles and specialist raptors.

Then there was possibly the most eccentric of all – Frances Hamerstrom – who lived and breathed Marsh Hawks (now Northern Harriers), on the grasslands and prairie chicken booming grounds of the Buena Vista Marsh in Wisconsin. She raised harriers, housed interns (gaboons) as research assistants to study them and uncovered the harrier's remarkable population cyclicity with its main prey, the lowly Meadow Vole. She did so by turning over hay bales each summer to count the escaping voles, one of the odder and most straightforward forms of small mammal census. In between, she revealed the depressing effects of DDT on the breeding success and mating patterns of her harrier population. In so doing she helped seal the fate of DDT in North America, by demonstrating its role in Northern Harrier and Peregrine Falcon population declines.

Far away down under, David Baker-Gabb leads the Aussie Harrier eccentrics and he suggested audaciously (and correctly), that the harrier genus *Circus*, arose in the southern rather than the northern hemisphere. Today, DNA studies now support his suggestion.

So it is little surprise that Don Scott, who is Irish, joins the ranks of those of us with a passion for these poorly-appreciated raptors and is among the most dogged and world-travelled of harrier aficionados. He is famed for his discovery that Hen Harriers in Northern Ireland regularly nest in trees, a habit shared only by Australia's Spotted Harriers. Don has taken on the task of describing and reviewing all of the world's 16 species of harrier, to make them more of a household name and to highlight their plight. He has managed to observe all 16 species and finally our dream of finding and capturing the world's most elusive harrier in Papua New Guinea, has been realised! This bird, like so many of the world's harriers, had simply been subsumed as a subspecies of the European Marsh Harrier, or the Australasian Swamp Harrier.

It is stark fact, that three of the world's 15 recognised harriers – Black Harrier, Madagascar Harrier and Reunion Harrier – are globally threatened species under IUCN Red Data criteria. At 20% of the group, this is higher than many other genera of other raptors and threatened birds. As ground-nesting species, harriers are susceptible on islands to introduced mammals such as cats and rats, and elsewhere they are losing ground to wetland drainage, pesticides and the degradation of foraging lands for agriculture.

It is essential, therefore, that the new generation of harrier torch-bearers (Steve

Redpath, Beatrice Arroyo, Simon Thirgood, Odette Curtis and Igor Fefelov, to name but a few), not only add to our knowledge of harrier ecology and behaviour, but use it to protect species that are Red-Listed and keep off that list, those that are not.

I am delighted that Don has taken this opportunity to breathe life into these birds and bring them more into the public eye. It is my hope that a book like this will help educate those who can make a difference to keep the world's harriers, as a suitable icon of the wild steppes and wetlands.

Rob Simmons
Cape Town, South Africa

Fifty years ago, I had the thrill of seeing my first two harriers ever; a Marsh Harrier, quartering low over a Norfolk reedbed, then, later in the day the beautiful male Montagu's Harrier in spectacular flight over another Norfolk Broad. I have never forgotten that day and have been fascinated ever since.

Don Scott, was also entranced by his first sight of harriers; Hen Harriers in the heaths and pines of Northern Ireland. For me, I have retained a great interest in harriers, but they are much more for Don; they have become his life's passion.

Almost alone, he has set out to learn about the world's harriers and to answer some of the many questions about them. At the latest count, there are 16 species and one recognised subspecies and in May 2007, Don completed his quest to see them all in the wild. Some were relatively easy to find but others in India, South America, Madagascar and remote Reunion Island needed considerable expeditions. When I last saw him, there remained the Papuan Harrier and he was weighing up the prospects of a trip to Papua New Guinea, a country which presents more hazards to a visitor than the purely ornithological ones. Now he has been there and succeeded, returning healthy and unharmed with some priceless observations on New Guinea's little known harrier. It has been a marathon of determination and enthusiasm and I doubt if there is anyone else who has had first hand knowledge of every one of the world's harrier species.

Harriers need a champion. They have never enjoyed the celebrity status of eagles or falcons and apart from a few species, remain very little known. In today's world, this is a dangerous situation. Man is forever encroaching on nature and nobody can plan for a species' conservation without knowledge of its requirements. Who could have foreseen, as Don has shown, that Northern Ireland's Hen Harriers would become dependent on plantations of pines, established purely as a source of timber? Many of the species need marshes and heathlands, both habitats which are increasingly threatened all over the world, but harriers seem unlikely to be considered in any development plans for such places.

In 2006, Don came to Australia and I spent a highly enjoyable time in the field with him. His knowledge of harriers is immense and his enthusiasm all consuming. His book promises to provide a wealth of new knowledge, to be a powerful weapon in the defence of harriers and to give great pleasure to all those who read it.

David Hollands
Orbost, Victoria, Australia

PROLOGUE

My fascination and passion for diurnal birds of prey and their closest congeners, the nocturnal owls, stems back to my youth and luckily for me it has remained into adulthood. This book is therefore the culmination of a 22-year obsession with a family of raptors, known as harriers. Experts say there is at least one book lingering in every person. I have chosen to dedicate it to these beautiful creatures which have given me so much pleasure for well over two decades and hopefully at least two more in the near future.

When I first began studying Hen Harriers, in my native Northern Ireland back in 1986, I had no idols or mentors and my only sources of information were C Douglas Deane (Jimmy) from Belfast and my namesake, the late David Scott from Dublin. Both these gentlemen had a passion for raptorial species and in the case of David Scott it was strictly harriers in the Wicklow Hills, a short distance from his home, whereas Jimmy Deane mainly preferred to photograph them in their natural habitat in the Glens of Antrim. I paid several visits to David Scott's home and have kept all his typed correspondence. He was small in stature, but had a big heart and was so passionate about harriers! It was David who rightly suggested I keep a diary or notebook of my harrier activities and this advice was gratefully accepted and at the time of writing I had already recorded almost everything that is mentioned in this book. When my son Douglas was born in 1989, he was intentionally named after C D Deane, whom I greatly respected and spent many hours in the field with.

The passing of these two accomplished ornithologists, in 1992 and 1998, placed me on a steep and fast learning curve on all aspects of the Hen Harrier's complicated and contrasting lifestyle in the counties of Antrim (my main study area) and Tyrone. Lacking financial and moral support for my work locally, I persevered and was soon self-taught in the art of studying firstly, Hen and then all the world's species of harrier, which is something I am extremely proud of. Luckily for me the UK-based Hawk and Owl Trust, financially supported my work here and abroad for many years, for which I am most appreciative. In September 2001 I was awarded a degree in Biology, by the Institute of Biology in London for my original studies of Hen Harrier in Northern Ireland. Then in January 2007, the title of Visiting Research Associate was conferred on me by the School of Biological Sciences, Queen's University, Belfast. I am also a founder member (1992), of the Northern Ireland Raptor Study Group.

Following mainly casual observations between 1986 and 1990, my scientific studies of Hen Harriers and the remaining 15 species and one subspecies worldwide, commenced in 1991. At that time, the only harrier reading material available to me was Donald Watson's impressive monograph of *The Hen Harrier*, published by Poyser. This was without question my first book purchase and a useful reference guide, until others were published in subsequent years by Hamerstrom, Clarke (3), Simmons and most recently Verma (2007), plus an early 1923 contribution, by the Danish author Henning Weis. All these fine books contain a plethora of useful information about the enigmatic behaviour and ecology of harriers, and all sit proudly on my bookshelf. Over the past

30 years or more, hundreds of well-written scientific papers on the now extended harrier family have been published in journals and bird reports all over the world. Consequently my reference library has swollen ten-fold to accommodate quite a few of them, along with other superb books on raptors and owls!

I have been so fortunate, in that my 'Harrier Journeys' have taken me to some wonderful and exotic locations around the world and to both north and south of the equator in my quest to find and study all the known species of *Circus*! My book could be best described as a calendar or a diary of events over a 22-year period where I encountered all of the world's harrier species (16) and one recognised subspecies, in their unique breeding and wintering habitats, after visiting over 20 countries to fulfil a long-standing ambition. Consequently, this book was specifically written not to confuse people with pages and pages of scientific jargon. I hope therefore that all those who read it, will find it informative, easy to digest and an insight into the world of harriers.

The one main difference between my book and the previous eight is that I have personally observed and studied all of the world's harrier species. Therefore, my regular correspondence with several of the world's harrier experts confirms that this feat has probably never been achieved before by those who habitually study members of the genus – *Circus*! As a result, a considerable amount of information that I have gleaned will not have been published before, hopefully making this book of value to those who are interested in this group of birds.

If I am not out in the field studying harriers at home or abroad, I have dedicated quite a lot of my spare time over the last 10 years, to displaying my captive-bred Great Grey, Barn and Eagle Owls, which are solely used for education purposes at WWT Castle Espie and in schools across Northern Ireland. To talk to children (and adults) of all ages about raptors in general, and of their rarity and beauty, and their need for urgent conservation throughout the world has been a worthwhile experience for me also.

ACKNOWLEDGEMENTS

I must admit that this book could not have been written without the help and unstinting support of many people, not only here in my native Northern Ireland, but in the various countries throughout the world that I visited to achieve my goal of studying the world's 16 species of harrier and the one recognised subspecies. From the start I knew my task would not be easy, but showing true Irish grit and determination finally saw me pass the imaginary winning post in Papua New Guinea, on Wednesday 2nd May 2007. During the course of my quest I met many old friends and thankfully lots of new ones along the way, many of whom I still keep in touch with. Thank heavens for email, what would we do without it?

Firstly, and most importantly, I would like to thank my long-suffering wife Linda and my son Douglas for their continuing support and encouragement to not only see the project through, but also to make sure that my 'Harrier Journeys', over the past 22 years, were finally accomplished by publishing them in book form. Linda and Douglas have been absolutely superb, especially when I vanished for days or weeks on end to some far-off country where harriers abound. I must also include my late parents William and Margaret, who not only gave me a steadying hand when required, but also their financial support – God bless them both!

During my extensive Hen Harrier studies in Northern Ireland, which have to date consumed 22 years, my constant companion through good times and bad has been my close friend Philip McHaffie. He has not only been a confidant of mine, but a true birding friend in every sense and I thank him sincerely for his outstanding and unselfish contribution. I must also thank another birding colleague Gary Wilkinson, who for his sins introduced me to Hen Harriers in 1986; little did he know then, how things would progress in later years. Larry Toal and I have been raptor colleagues for over 20 years and I hold him in the highest regard. Thanks Larry for all your help each season with unlimited Hen Harrier information.

In the Forest Service I extend my thanks to Basil Lenaghan for showing me the albino Hen Harrier in County Fermanagh and to John Mooney for looking after my harrier interests in County Tyrone; not forgetting Fred Quinn who keeps me up to date with harrier activities in County Antrim. A big thank you also goes to Stanley Black, estate manager at Cleggan Lodge, for granting me access to the area during my ongoing studies. Local photographer Derek Charles, kindly donated several stunning images of Northern Ireland's second Montagu's Harrier.

I could not leave out the late John McGhie (ex Forest Service) and the late Dr William McDowell, who both died far too young in 2005 and 2006. Both took the time, when they were probably busy with their own commitments to help me, firstly with the conservation of breeding and roosting Hen Harriers in the forests of Counties Antrim and Tyrone and secondly, with positive comments and advice when proof-reading several papers I hoped to publish. Both were a credit to conservation and birding in Northern Ireland. To George Gordon (Records Secretary of the Northern Ireland Birdwatchers' Association), who was always willing to help me when I required that little bit more information about harrier records in Northern Ireland – thank you George for all your help over the years.

In the neighbouring Republic of Ireland I hail my good friend Lorcan O'Toole. Lorcan has been a tireless and inspirational raptor worker in the Republic for many years, which I for one have greatly admired and supported and I thank him for his most welcome Hen and Marsh Harrier information. Dr Stephen Newton from BirdWatch Ireland, has also been supportive of my harrier work and as editor of *Irish Birds* has kindly published and prepared final drafts of my papers – many thanks Stephen. I have corresponded with Frank King from County Kerry for two decades regarding Hen Harriers. Sadly we have never met in the field, due to the long distance we live from each other. Hopefully this can be achieved in the near future. When I visited the nearby Isle of Man, I was firstly made welcome by Dr John Thorpe and then later by my good friends Geoff and Judy Kelly, to whom I am most grateful for their unstinting hospitality and kindness.

In England, I have to be thankful for the help provided to me by harrier expert, the late Dr Roger Clarke and Colin Shawyer, who in the first instance supported me when others in my own country failed me, particularly when it came to proving that Hen Harriers in County Antrim really did nest in trees. This support has thankfully continued for the past 17 years, but was brought to a premature end by Roger's death, on 28th January 2007, but still continues through my 20 plus years of unbroken friendship with Colin and his wife Val. My three trips to India and one to Holland, with Roger were momentous towards my understanding of harriers worldwide. I must freely admit that I learned quite a lot about *Circus* species, thanks to Roger's influence. Graeme Hewson and David Miller from England were fantastic company when they joined Roger and I during our 1999 visit to the subcontinent. I also thank Nigel Middleton for providing me with moulted Pied Harrier feathers for DNA purposes. Thanks also to Arthur Grosset from Suffolk and Peter Nash from Swinton, for sending me their excellent photographs of Long-winged and Cinereous Harriers, and to Mike Wilkes from Redditch and George Reszeter from Abingdon who kindly provided superb pictures of Montagu's Harriers.

In Scotland, I must thank Brian Etheridge for all his help and useful advice which was always instantly given to me when required and both he and Ricky Gladwell, always made me feel welcome when it came to sharing harriers, in their country. The late Don MacCaskill was also a close ally of mine for many years when it came to discussing Hen Harriers. Mike Gregory from Argyll, Bob McMillan from Skye and Don Smith from Ayrshire, have all been extremely helpful to me regarding photographs and ongoing information on harriers and other raptor species. Not forgetting the late Donald Watson, who at the age of 73 and with failing eyesight, bravely came and supported my tree-nesting find and also wrote a follow-up paper to that effect. I also thank Bill Bourne from Aberdeen, for being supportive of my work and also for publishing his thoughts on tree-nesting. An inspiration to me for many years regarding raptors has been the incomparable Roy Dennis – thanks Roy for your words of wisdom over the years that I have known you! A superb photograph of a juvenile Pallid Harrier has been provided by Hugh Harrop from

Shetland, and from Steve Redpath at Aberdeen University a unique albino Hen Harrier image. Many thanks to you both.

In Wales, I sincerely thank the talents of bird artist Philip Snow, whose paintings and drawings of harrier species grace this book. When speaking with Philip many years ago after viewing his exceptional work in other bird books and contributions he made to me personally, I made a promise to him then that if I ever committed to writing a book on harriers, he would be my chosen artist. I am so glad he has agreed. Although he has now moved on to mainly film and TV work, Iolo Williams and I corresponded and swapped information on our respective Hen Harrier studies which was most interesting and has proved helpful to the both of us.

When my studies of both Hen and Western Marsh Harriers moved to Europe, my friends Henk and Marlies Castelijns from Holland, always made their home available to me and I thank them both sincerely for their hospitality at all times. I also add, Andre Bourgonje, who without fail would travel from The Hague to meet me and participate in my harrier activities, at Saeftinghe. Old friends, Jean Maebe, Marc Ploegaert and Jaap Poortvliet, also provided me with many memorable harrier moments. Superb photographs of Western Marsh Harrier have been kindly donated by Ludo Goosens and Niels De Schipper.

Also in Europe, I thank Pere Vicens at the Parc Natural De S'Albufera in Mallorca, who took the time to discuss with me numerous queries about the resident subspecies of the North African race of the Western Marsh Harrier – *Harterti,* and for showing me the island's first ever breeding Montagu's Harriers. During subsequent visits I was always made welcome at this world renowned reserve. Sincere thanks also go to my Russian harrier colleague Igor Fefelov and Bjorn Johansson from Sweden, whose information and photographs of Eastern Marsh Harrier have been most helpful for this book.

Moving worldwide, I unreservedly thank Dr Robert Simmons and Dr Andrew Jenkins, at the Percy Fitzpatrick Institute of Ornithology in Cape Town, South Africa, for all their help and guidance during my studies of breeding Black and African Marsh Harriers, in 2002. My continuing friendship with Rob, whom I personally regard as one of the world's leading authorities on harriers – his work on this Genus, has been inspirational to me for many years. I also thank him for taking the time to write the *Foreword* for this book and also providing me with beautiful photographs of Black and Papuan Harrier. Andrew Jenkins has also provided me with superb photographs of the Black Harrier – many thanks Andrew! Raptor expert, Peter Steyn, kindly made available much needed photographs of African Marsh Harrier – thank you Peter! During my many visits to the Gambia, the Birdman, the late Mass Cham, who sadly passed away on 13th June 2006, was not only a dear friend, he was also a man who new his raptors and other bird species. Our trips into the interior of that small country to seek out raptors were always filled with action and his knowledge and advice were always greatly appreciated – you are greatly missed Mass!

Thanks to Kirsty and Allan Blanchard from Tiger Trails, who organised my trip to Assam in 2003 and also to Maan Barua and his family at the Wild Grass Resort, in

Kaziranga National Park, for their hospitality and birding information concerning Pied and Eastern Marsh Harriers. Also in India, Rishad Naoroji from Mumbai has remained a good friend and a regular correspondent for the past decade – it is always good to hear from him and his unceasing raptor activities in that great country. I was absolutely delighted when Rishad decided to come to Northern Ireland and meet me on 4th September 2007, during his punishing vacation to Europe. Vibhu Prakash, a raptor colleague from the Bombay Natural History Society, was great company during my 1997 visit to India. To Sarfraz Malik and family from Dasada, who kindly looked after me during my 1999 visit to India in a traditional Indian homestead. I thank them unreservedly for their unique hospitality. Manoj Pai, from Ahmedabad has also been a great help to me with harrier information, with Jugal Tiwari from Kutch and Ashok Mashru from Rajkot, kindly assisting me with their photographs of Pallid and Western Marsh Harrier. I am also grateful to another harrier colleague, Ashok Verma from Dehradun, for generously forwarding me a copy of his recent book – *Harriers in India: A Field Guide*.

Also in Asia, I thank bird photographers, Lim Kim Chye from Malaysia, Romy Ocon from the Philippines, Mervin Quah from Singapore and John and Jemi Holmes from Hong Kong (esp. Jean Hosking FLPA-Images), for providing me with stunning photographs of Pied and Eastern Marsh Harrier, which now complement this book.

On my way to Reunion Island in 2004, to observe the endemic Reunion Harrier, I had the good fortune to meet Peter and Nancy Vajtai and their daughter Beverly, at Charles De Gaulle Airport, Paris, during a time of confusion and uncertainty for everyone travelling to Reunion. Their company at this time was most welcome and is still appreciated today. In Reunion, I have to thank, Francois-Xavier Couzi and Marc Salamolard from SEOR (Societe d'Etudes Ornithologiques de la Reunion) and Laurent Brillard for all their vital information and photographs of the Reunion Harrier. To Derek Schuurman from Rainbow Tours, who did a great job of organising my joint visit to Reunion and Madagascar – many thanks Derek!

On the island of Madagascar my thanks entirely go to three people. Firstly, Russell Thorstrom from the Peregrine Fund (USA), who sent me study slides and papers regarding the Madagascar Marsh Harrier, which proved to be invaluable during my visit and secondly, Lily-Arison Rene de Roland who provided me with wonderful photographs for this book. Thirdly, my thanks go to the lovely Jocelyne Randimbison, who diligently looked after me during my visit to Ambohitantely, in the Central Highlands. In all my travels, I had never acquired a female guide before – she was brilliant having to put up with me!

In December 2004, I had the pleasure of meeting my most able guide Sergio Corbet, in Buenos Aires, Argentina. We spent an incredible two weeks together, firstly studying Cinereous Harriers in the steppes of Patagonia and then in the Pampas with Long-winged Harriers. His expertise at finding both species for me to study in good numbers was priceless – muchas gracias Sergio! Thanks also to my hosts in Patagonia, Nancy and Ludie Henning and also to Natalie Collm, who kindly provided me with superb pictures of Cinereous Harriers, to use in this book.

When I visited North America for the first time in January 2006, to study wintering Northern Harriers, my amigo, Sergio Corbet from Argentina, recommended Josh Larsen from Florida. What a good choice, and thanks to Josh and his partner Yraima Hernandez my goal was achieved. Josh also provided me with superb flight shots of Long-winged Harriers, to use in this book. Thanks again Sergio, I owe you one; after all you did introduce me to Quilmes, the best beer in Argentina! Thanks also go to Dorn Whitmore, Marc Epstein, Jim Lyon and ranger, Nancy Corona at Merritt Island NWR, Florida, for all their help and support, and in particular for allowing me unlimited access to the Refuge before sunrise and after sunset in my pursuit of roosting Northern Harriers.

I first met Bill Clark from Texas, in 1993 at a raptor conference at the University of Kent and then again in 1997, while with others studying wintering harriers in India. Bill has always been most helpful to me over the years that I have known him and I thank him sincerely for allowing me to use several of his photographs in this book. I pay homage to that Grand Old Lady of Harriers – Frances Hamerstrom, who studied Northern Harriers in the prairies of Wisconsin from 1959 until her death in 1998, aged 91. I first met her in 1993 at the same raptor conference in Kent and her enthusiasm for these birds even then was so infectious and belied her 86 years. During our most interesting conversation she proceeded to sign her famous harrier book for me. The inscription reads, 'For Don Scott, – world authority on tree-nesting harriers'! What a lady! Thanks also to George Jameson from Washington State, for his superb picture of a male Northern Harrier.

During my visit to Australia in October 2006, to study breeding Swamp and Spotted Harriers in Victoria and New South Wales, several people went out of their way to ensure that I had a highly successful trip and are worthy of more than just a brief mention. They are Drs David and Margaret Hollands, who made me so welcome, not only to Australia, but also to their homes in Melbourne and Orbost. Special thanks to David for agreeing to write the *Preface* for this book, and also for his positive comments on the complete text and for giving me free access to use his excellent photographs of Spotted, Swamp and Cinereous Harriers.

A big thank you also goes to Australian harrier expert Dr David Baker-Gabb, who has corresponded with me since 1994 and over the years has kindly forwarded several of his papers on both species for my personal use. It was great to meet up with him in Melbourne prior to going home. In New South Wales, three people stand out immediately to receive my personal thanks. First and foremost Gwennie McLaughlin and Murray Creed, who kindly looked after me irrespective of Murray's illness – (get well soon Murray!) If it had not been for David Webb, my task of finding a breeding pair of Spotted Harriers in NSW would probably not have materialised due to the severe drought that had badly affected this and other parts of Australia in 2006. On two separate occasions David travelled a considerable distance from Griffith, to show me the only known nest in the area of Urana, and also provided me with superb photographs of Spotted Harriers, to use in this book. Generous thanks also go to Graeme Chapman from Queensland, for his in depth photographs of a Spotted Harrier's nest.

The last leg of my 'Harrier Journeys', was to Papua New Guinea in May 2007, to study the little known Papuan Harrier, so there are quite a few people I must thank personally for making this trip a complete success. It is highly unusual to acknowledge one person twice in the same book, but in the case of Dr Rob Simmons, I make an exception. Were it not for his sheer persistence in acquiring and then eventually securing the necessary funding, it would not have been possible for me to participate in this ground-breaking expedition. Only he, made it possible, otherwise I would probably have never fulfilled my dream of studying and writing about the world's harrier species. To John and Michael Simmons, I thank them for their friendship during what was an exciting and momentous occasion for everyone.

Special thanks go to our guide Leo Legra and driver Onika Okena, from the Wildlife Conservation Society in Goroka; they were always available without question, every day. Were it not for the exceptional help provided by Rex Topiso, the regional airport manager, at Mt Hagen and electrical supervisor, Russel Palia, our initial studies of this species would never have materialised. They deserve more than just a big thank you, as does TK, the airport manager at Goroka, for allowing us unlimited access to both these airports. I also thank all the staff at the Highlander Hotel in Mt Hagen and Kumul Lodge, in Enga Province, in particular our bird guide Max Poliaka. In Goroka, I have to thank unreservedly, Brendan and Pippa Ellis, the owners of Goroka Preparatory School, for their unstinting hospitality during our unscheduled stay there; they were fantastic hosts as were the staff, teachers and not forgetting the 40 lovely children, who all made us most welcome. To driver David Segeye, Kimmy, Oscar, Mike and Yombu from Yabiufa Village, a hearty thank you goes to them all. Adam Buke from Air Niugini also receives my sincere gratitude.

I personally thank my main sponsors namely, Hawk Mountain Sanctuary, USA (especially Keith Bildstein), Professor Michael Wink (Heidelberg University, Germany) and John Simmons (England). I also thank James Orr, Director of WWT Castle Espie NI, and Julia Carson of the Viridian Group NI, for their kind gesture of T-shirts for me to wear during my trip to Papua New Guinea.

In conclusion, to lose five birding friends, all of whom have been acknowledged here, in less than two years has been a sad time for me personally. Their talents will be greatly missed throughout the world of conservation and birding. Two of the five, were huge stalwarts from the small band of people who religiously study *Circus* species worldwide. Their absence from the birding scene (as the others) has left a huge void, which will not be easy to fill.

Finally, I must thank Tiercel Publishing, and in particular Colin and Val Shawyer, for without their help and understanding, this book would probably have never gone to print.

If I have inadvertently left anyone out, I humbly apologise. I thank everyone near and far who may not have been personally mentioned. Over the past two decades, I have spent thousands of pounds, travelled thousands of miles, but more importantly, shared thousands of hours with my favourite birds of prey – the harriers!

Don Scott
Dundonald, Northern Ireland

Portrait of an African Marsh Harrier in Africa

Peter Steyn

A handsome male Hen Harrier – pale grey with a snowy-white rump and black-tipped wings was quartering the moorland, seductively lead by his mate, who despite her dull plumage, lured him to a chosen spot in the heather, that would be their temporary home for the next three months. So you may well ask, what is so exciting about harriers?

For the uninitiated, harriers *circus spp*, are medium-sized, elegant, long-winged, long-tailed and long-legged members of the hawk family *accipitridae* which mainly inhabit open country, hunting low over the ground and reedbeds, with slow methodical wing-beats, alternating with short glides on slightly raised dihedral 'V' shape wings. They have a flatish face, with an owl-like facial disk and for most species, prominent white uppertail coverts, which are only visible in flight. Juveniles and most adult females have dark brown upperparts, streaked underparts, mainly rufous on the former, with banded tails and because of this, are commonly known as ringtails. The word ringtail has been formally used for centuries and was first motivated by the bars (rings), on the tails of female harriers. Most male harriers tend to have much paler plumage details and are generally of smaller size and much lighter in weight than females. The smallest and lightest of all the harriers is the Montagu's Harrier from Europe, the largest, the Long-winged Harrier from South America and the heaviest, the Swamp Harrier from Australia.

Harriers have several traits that separate them from many other raptor species, one of which is the spectacular aerial transfer of prey – the 'food-pass', and the pre-breeding courtship display known as 'sky-dancing'. Nests are generally found on the ground amongst deep heather, reeds and other rank vegetation. Until 1991, only one species, the Spotted Harrier from Australia, habitually nested in trees, after which a pair of Hen Harriers followed suit in a mature conifer plantation in County Antrim. Northern Ireland remains the only country throughout the Hen Harrier's extensive European breeding range where this

Western Marsh Harrier food-pass
Philip Snow

The stunning Spotted Harrier of Australia

David Webb

still occurs. Another trait that distinguishes harriers from many other raptor species is polygyny. It is known to occur with regularity in at least four harrier species, rarely in one, highly suspected in another and probably as yet, undetected in others.

Harriers are regarded as highly skilled and opportunistic predators choosing to hunt at any time of the day, when weather conditions are favourable, when prey is most active or when other raptors are not hunting.

Outside the breeding season, communal roosting often occurs in a variety of ground habitat types and is another spectacle which can involve as few as one bird, as many as 50, or even in excess of 100 (common in the Isle of Man), to over 3,000 at one particular site in India. Other less common forms of roosting have been recorded in tall coniferous trees, on bare open ground, on the tops of hedges and even in vacated tree nests. The latter occurs more commonly in Hen Harriers which are wintering in Northern Ireland. Harriers regularly regurgitate a small pellet (or a cast) of indigestible bones, fur and feathers and when deposited at nests and roost sites, they can be analysed allowing us to determine the components of their diet at different times of the year.

Despite being attractive and exciting birds to observe in the field, harriers worldwide are regarded as extremely rare and vulnerable throughout most of their range, due to deliberate disturbance, loss of habitat, predation and most worryingly, persecution. At present, 16 species and one subspecies are recognised throughout various parts of the world, from as far apart as North America to New Zealand.

The word 'harrier' is derived from the Old English *herigan,* meaning to harass by hostile attacks. *Circus* is from the Greek *kirkos,* meaning a bird of prey that flies in circles. The collective noun for birds such as harriers – is a swarm.

Why are there 16 species and where are they found? Thirty-four years ago in 1973, the respected Dutch taxonomist Ebel Nieboer suggested that there were 10 species of harrier worldwide – from Europe and eastern Russia, the Hen, Western Marsh, Montagu's and Pallid; from the continent of Africa, the African Marsh and Black Harrier; from Siberia and Mongolia, the striking Pied Harrier; from Australia, the then only habitually tree-nesting species, the Spotted Harrier and from South America, the Cinereous and Long-winged Harriers.

Fifteen years later in 1988, three new species were added by American researchers Dean Amadon and John Bull. They eventually recognised and gave full species status to the Madagascar and Reunion Harriers (considered then to be a single species) and to the two Marsh Harriers of the Eastern Hemisphere – the Eastern and the Swamp.

Twelve years on, in 2000, Professor Michael Wink and his colleagues in the genetics laboratories of Heidelberg University in Germany, along with harrier expert Dr Rob Simmons from South Africa, and researchers from around the world, who donated feathers and blood samples from which DNA was successfully extracted, now suggest the addition of a further three species. This now increases the number of species worldwide to 16, with 13 in the Old World and three in the New World.

These three new species are the Reunion Harrier, which differs genetically and morphologically from the much larger and heavier Madagascar Harrier, with males and females of the former being much darker in overall plumage detail. In North America, the Northern Harrier (formerly the Marsh Hawk), has finally and rightly so, been split from the European Hen Harrier, of which it was formerly regarded as a subspecies. The third species (but only proposed at present for full status), is the little known Papuan Harrier from the remote island of New Guinea; it is proposed to split this species from its nearest congener, the Eastern Marsh Harrier. The Papuan Harrier is considered to be a prime species for full status as it is known to be sedentary on New Guinea and is apparently morphologically smaller. In the past, island forms of harriers, have all proven to be excellent subjects for the purposes of genetic testing. So to all those "Harrierphiles" out there – watch this space! Finally, the subspecies – *Harterti,* is found sparingly in North Africa, with birds occurring in Algeria, Morocco and Tunisia and occasionally in southern Europe.

The harrier species I travelled the world to see are now described in the forthcoming chapters in the order I observed and studied them.

Harriers of the World

Common Name	Latin Name	Distribution
Hen Harrier	*Circus cyaneus*	northern Europe, Russia
Western Marsh Harrier	*Circus aeruginosus*	Europe, western Russian
	C.a.harterti	North Africa
Montagu's Harrier	*Circus pygargus*	Europe, western Russia
Pallid Harrier	*Circus macrourus*	Russia, eastern Europe
Black Harrier	*Circus maurus*	South Africa, Namibia
African Marsh Harrier	*Circus ranivorus*	southern central Africa
Pied Harrier	*Circus melanoleucos*	Mongolia, China, Myanmar
Eastern Marsh Harrier	*Circus spilonotus*	Asia
Reunion Harrier	*Circus maillardi*	Reunion Island
Madagascar Marsh Harrier	*Circus macrosceles*	Madagascar, Comores
Cinereous Harrier	*Circus cinereus*	South America
Long-winged Harrier	*Circus buffoni*	southern central America
Northern Harrier	*Circus hudsonius*	North America
Swamp Harrier	*Circus approximans*	Australia and New Zealand
Spotted Harrier	*Circus assimilis*	Australia
Papuan Harrier	*Circus spilothorax*	New Guinea

European Harriers – male and female   *Philip Snow*

Spotted Harrier carrying prey

David Hollands

Places I visited on my journeys

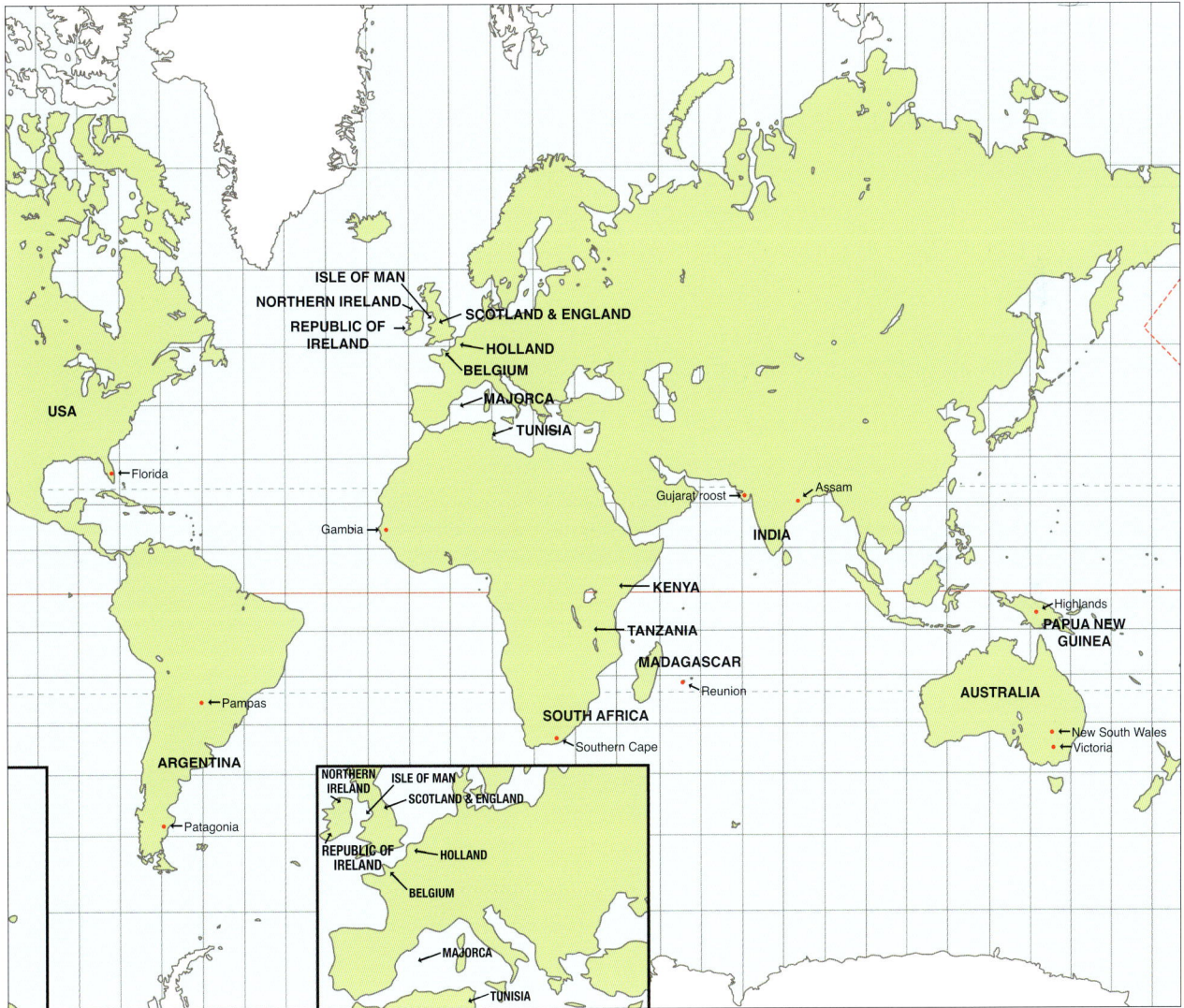

Adult male Hen Harrier

2

HEN HARRIER

Circus cyaneus

The **Hen Harrier** has not only the most northerly breeding range of the Western Palaearctic's four species of harrier; it also has the widest distribution, stretching from western Europe to eastern Asia. Birds breed from as far west as Ireland, the UK, and in at least 16 countries of continental Europe, and then east to Kamchatka and Sakhalin. It is the most studied of all 16 harrier species and also the most persecuted, especially in the UK uplands. Love them or hate them, no other raptor has caused greater controversy throughout the United Kingdom than the Hen Harrier, due to its liking for game birds, and its ongoing conflict with the shooting fraternity.

In his *Birds of Northern Ireland*, published in 1954, the late great C D (Jimmy) Deane, described the Hen Harrier as a rare vagrant, which had only been recorded four times since 1900, yet a century earlier it had bred commonly in five of our six counties, the exception being Armagh.

Then in the 1960s, following extensive planting of nursery plantations (of young conifers), on the heather clad slopes of Northern Ireland, especially in County Antrim, the Hen Harrier began to thrive again in our uplands. It selectively placed its nest at the base of 1m high conifers, where the site was wide open on all sides. By the mid 1970s, some 20 pairs were breeding in County Antrim alone, but when the trees grew taller and the canopy began to close, the harriers deserted the Spruce plantations and returned to the open moorland. In the early 1980s, numbers began to decrease again, probably due to over-grazing by numerous sheep and widespread predation from foxes. After 22 years of continuous studies of this species, only in 1999, did my colleague Philip McHaffie and I ever encounter 20 plus pairs of Hen Harriers in the Antrim Hills. This was almost certainly due to a dramatic vole crash in south and east Ayrshire and likewise in Dumfries and Galloway, which also affected the Barn and Short-eared Owl populations there. The years preceding 1999 have been a catalogue of neglect, decline and mismanagement of our uplands, in which none of our conservation organisations have come out of it bathed in glory!

It was during the spring of 1986 that I was introduced to my first ever pair of Hen Harriers, by my good friend and fellow birder, Gary Wilkinson, at Slieveanorra Forest, in County Antrim. I was immediately captivated by the harriers' graceful flight and also their sheer beauty and elegance, as they performed breath-taking displays above the forest canopy and the surrounding moorland. The apparent rarity of this harrier and my lack of interest in non-raptor species and twitching, gave me the necessary impetus to commence my studies, that year. This species of bird, after only one season had an early influence and immediately shaped my ornithological career, the rest is history!

I have had a long and interesting association, spanning over two decades with this species, since observing my first 'sky-dancing' pair in the Antrim Plateau, in 1986. After strictly casual observations (1986-1990), I commenced my scientific studies of this elegant and much maligned raptor in 1991. It was then that my first ground and tree nests were visited for the purposes of gathering breeding data, and pellets were also collected to assess the bird's dietary requirements. Monitoring was carried out under licence from the DOENI and as a member of the Northern Ireland Raptor Study Group.

The diet of Hen Harriers in Northern Ireland is not as diverse as in other parts of the UK. The island of Ireland as a whole has never supported a large Hen Harrier population, mainly due to the absence of the Field Vole. Voles are also absent from the nearby Isle of Man, yet it supports a substantial and productive population of Hen Harriers, with around 40 breeding pairs annually and record numbers of wintering birds.

Over a five-year period (1991-1995), a total of 160 pellets were collected from ground and tree nest sites in County Antrim, my main study area, and from a winter roost site in County Tyrone. All the pellets were sent to Roger Clarke in England for analysis and the findings were recorded on an annual basis. The results showed that three species of small birds dominated 74% of the Hen Harrier's food supply; these being Meadow Pipit (32%), Starling (26%), and Skylark (16%). In total, avian prey made up 79% of their diet, with Lagomorphs (rabbit/hare), small mammals and unidentified beetle remains, making up the remaining 21% of their daily diet. Not surprisingly, Red Grouse did not feature highly (1%), given that there are only small pockets of this rare gamebird in the Antrim Hills and the Sperrins. In Scotland and other upland areas of the UK, when the Hen Harrier chicks are close to fledging, their parents prey more heavily on Red Grouse, which, in recent years, has rendered the harriers vulnerable to illegal persecution. In England today (2006), the Hen Harrier is regarded as a very rare breeding bird of prey, with only around 10-12 successful nests, annually.

A second batch of 310 pellets, collected over a ten-year period (1996-2005), from the above-mentioned counties in Northern Ireland, was analysed by myself, as I had built up an extensive reference collection, containing a wide variety of feathers, skeletal remains and samples of fur and hair from small and medium-sized mammals. As this sample included 150 more pellets and 193 more prey items, it soon became clear that Northern Ireland's Hen Harriers had a slightly more diverse diet than was first thought.

A further eight species of birds were identified and again Meadow Pipit, Starling and Skylark dominated 75% of the total diet. More significantly, Skylark numbers had decreased by over 2%, in line with the overall decrease of this species throughout the UK. Avian prey therefore accounted for 82% of the Hen Harrier's diet, and small and medium-sized mammals accounted for18%. The Viviparous or Common Lizard, appeared in pellets for the first time. One large pellet which was found below a tree nest, contained a large ball of hair (only), and when the contents were analysed by Roger, they strangely belonged to a Badger! A foraging Hen Harrier, probably the much larger female, had obviously found a Badger carcass and due to its thick coat and skin, the bird had been unable to successfully feed on it. The pellets also showed that Curlew chicks and Snipe were being taken by the harriers, as several carcases were also found on the forest floor, and clinging to the upper branches of four tree nests. Over a decade later, the Hen Harrier, the scourge of the moorland as far as the shooting fraternity is concerned, is still being persecuted and blamed for Red Grouse reductions in the Northern Ireland uplands, even though overwhelming evidence suggests otherwise!

Female Hen Harrier, hunting

Brian McGeough

Harriers are also great opportunists when it comes to exploiting certain areas for prey. During the 2000 breeding season I observed a grey male repeatedly flying to a precise area of moorland and then returning to a tree nest site within seconds with unidentified prey items. After its third visit I walked the short distance to where it was continually flying and discovered that it had raided a Meadow Pipit's nest and deliberately killed all the chicks. Having already brought three to the nest the remaining three were found lying side by side awaiting collection which occurred a short time later. Then, in 2005 a female flew 300m from her nest on three separate occasions over a 20-30 minute period (Howard Williams pers comm) to take full advantage of a nest of rats in a recently cut hay field at Belmore in County Fermanagh. It appears the prey became exposed after the field was harvested, with the female provisioning her rapidly growing brood on the spoils. Although my informant has only observed harriers obtaining prey on a few occasions in this area over a seven-year period, interestingly, the quarry has always been mammals.

Amazingly, on 14th July 1991, I discovered that Hen Harriers were tree-nesting on the deformed trunk of a 4.5m high Sitka Spruce, which had lost its leading shoot at an early stage of growth. This unique behaviour still occurs annually, with nests ranging in height from 2-13m high and is only to be found in Northern Ireland. Previously, the beautiful Spotted Harrier from Australia was the only member of the genus *Circus*, which habitually nested in trees. In both cases, poor ground cover and predation had forced these two species to take up residence in trees.

In Northern Ireland, not surprisingly, the find was treated in a very low-key way. Only welcome intervention from harrier expert Roger Clarke, and Colin Shawyer from the UK, helped me to bring this unique discovery to the notice of the birding fraternity and other harrier experts worldwide. So interesting was this find that Hen Harrier expert Donald Watson, travelled all the way from southwest Scotland to observe this unique nest on 23rd July. Donald had been studying this species for over 50 years and had also written a fine monograph of its breeding ecology in the Galloway Hills, but this was the first time he had encountered Hen Harriers, tree-nesting. One sceptic here, had the audacity to suggest that it was Common Kestrels I had observed and not to mention it to anyone for fear of causing personal embarrassment, how wrong that person was. Thankfully, my invitation to him to come and visit the nest site was not taken up.

This famous site was revisited in November 1991 and a small amount of fresh excreta was found below the long-vacated nest. When arboreal nesting unexpectedly occurred again in the same forest in 1992, both sites were simultaneously visited in October. Below both nests were substantial amounts of crystallized and recently deposited excreta, which could not have remained from the past or previous breeding seasons due to the harsh winters that persist in our uplands. In early February 1993, when revisiting the 1992 site, I inadvertently flushed a grey male, which instantly told me that individual Hen Harriers were utilising vacated tree nests, as roost sites during the winter months, another first for this amazing species!

My first taste of observing a large roost of Hen Harriers occurred on Saturday 2nd November 1991, when a local birder on the Isle of Man, Dr John Thorpe, invited Roger Clarke and I for a long weekend to view what he described at the time, as a considerable roost of Hen Harriers. I had only met Roger for the first time on 20th July, when he arrived in Northern Ireland with Colin Shawyer to confirm that Hen Harriers had actually tree-nested for the first time, so this was another chance to meet up again.

We arrived at the site in the Ballaugh Curraghs, which is in the northwest of the island, around 2.00pm and met up with another harrier enthusiast, Geoff Kelly. Owing to the height of the surrounding hedges and bushes, Geoff had skilfully built a platform on top of a trailer which he transported from his nearby home to view the roost. If this platform had not been available, we would not have been able to observe the area or the birds properly. The size of this site was incredible and to count the correct number of incoming birds, appeared almost impossible, owing to small bushes, willow-scrub, sedges and other unseen snags.

By 2.30pm we had commenced our watch, with the first harrier, a male, arriving at 2.41pm. After the first hour we had only recorded five. The second hour produced 33 birds and when we ceased at 5.15pm, a further 35 had settled at the roost. Our total for the evening was an amazing 73, of which 34 (47%) were grey males, the remaining 39 (53%) being females and ringtails. I was told that prior to my arrival over 100 had roosted the previous weekend, an incredible total for a UK roost site. Our second count on the 3rd, also commenced at 2.30pm and yielded a similar number (71), with only 28 (39%) grey males, 43 (61%) females and the occasional ringtail.

Our third and final watch did not commence until 3.00pm and when we called it a day at 5.05pm, we had amassed a total of 82 roosting harriers. By the end of the first hour we had recorded 18 birds, but the second hour produced an astonishing 63 harriers, and finally a single straggler to bolster our total even further. Of these, 31 (38%) were males and 51 (62%) were birds in brown plumage. If we had commenced our watch earlier we probably could have achieved 100 roosting harriers, that day.

By kind invitation of Geoff and Judy Kelly, I returned to the Isle of Man on 4th January 1992, for another three-day visit to this unique roost site. The peak months for roosts are generally October, November and December, and so I was not sure what numbers I might expect in January. Although numbers turned out to be somewhat reduced from my previous visit our first watch that evening produced a respectable 54 birds, of which only six (11%), were males and 48 (89%) were females and ringtails. This small number of males suggests that probably they had returned to their breeding grounds early (normal behaviour for grey males) only severe weather forcing them back to lower ground again.

My first ever morning watch on the 5th, had to be abandoned at 9.30am, due to torrential rain, with again six males observed and only 12 brown-plumaged birds. The afternoon count saw only three males arriving and 19 females. My second early morning watch on 6th produced nine males and 22 females including a few ringtails. Obviously six more males and three females had arrived much later the previous evening. My last evening watch on the 6th was more successful with 42 birds arriving, with again nine males (21%), and 33 (79%) brown birds. Counts at most winter roosts in Europe and the UK usually show an excess of females and ringtails over grey males, and this site was no exception. The contrast in numbers over a two-month period, from November 1992 to January 1993, was remarkable, with an overall decrease of 48%.

On 20th September 1992, a Hen Harrier winter roost site was rediscovered within the confines of a mature conifer plantation in the Sperrins, County Tyrone. This roost was in an unused turbary, formerly used by peat-cutters and was surrounded on all four sides by tall coniferous trees, with the harriers utilising an area of deep heather close to and directly behind four willow bushes, towards the rear of the site. Up to and including the 2005/06 winter roost survey, the site has been watched over 220 times, during a 14-year period, with a maximum of eight birds observed on 28th September 1997. In recent years, harriers have been in short supply, due to numerous intrusions by Red Foxes, which have forced the birds to occupy the surrounding conifers and most recently an alternative and as yet unfound site. One such intrusion, on 10th October 2001, saw a Red Fox, bobbing through the deep heather carrying a male Hen Harrier in its jaws. Away from the roost site, persecution of raptors and disturbance appears to be rife and this may account for the low numbers during watches with the majority of birds seen, tending to be adult grey males. Very few females, or young of the year are observed.

As predominantly ground-nesters and with most winter roost sites also on the ground, Hen Harriers are highly susceptible to predation from both avian and ground predators. Avian predators in Northern Ireland could include Common Buzzard or Goshawk, with the most common ground predator here being the Red Fox. Over the past two decades, I have personally observed several ground nests in County Antrim that have been totally ransacked by foxes and these incidents can be quite disturbing irrespective of how long one studies these birds. If a Red Fox is the culprit when the nest contains large young, the tell-tale signs include chewed pin feathers. Similarly, if the eggs have not hatched small pieces of egg-shell will be found in and around the nest site. On the other hand, if young are removed from nests at less than two weeks old and before the feathers come into pin, the removal is usually clean, with little or no evidence left for the observer to ponder over.

While monitoring 11 territories and eventually nine known nest sites on the Isle of Skye in 2007 (Bob McMillan pers comm.), only three of these were successful (33%), fledging 10 young, with the remaining six nests (67%), falling victim to marauding foxes. For those who religiously study this species, fox predation can have a devastating effect on small and vulnerable harrier populations like those in County Antrim and the Isle of Skye. Interestingly, and to my surprise, the Skye population is somewhat similar to that of the Antrim birds, as it is also heavily dependant on forestry plantations, even where these are up to 25 years old, with over-grazing and the burning of the moorland being the main reason.

A chance sighting of a lone grey male in early November 1994, which amazingly

A curious adult male Hen Harrier

Bob McMillan Skye-Birds.com

roosted in a tall stand of conifers in County Antrim's largest forest Slieveanorra, sparked interest in this area, but sadly it only lasted for two consecutive winters. During one watch on 27th November, six Hen Harriers, four males and two ringtails, roosted near the top of the 20m high conifers. In the following winter of 1995/96, five harriers utilised the tree-tops on 8th October 1995 and further visits to March 1996, revealed roosting harriers. This unique roost was linked to the winter movements of a flock of 2,000-plus Starlings, which also roosted in the forest close to the harriers. The large Starling roost also attracted several Peregrine Falcons, Common Kestrel, Sparrowhawk, and on one occasion an overwintering male Merlin which could be observed hunting on most evenings. By the start of the 1996/97 winter, the harriers, Starlings and the other predatory birds, had moved on and were not located in this area again, which sadly brought about the demise of this unusual roost site.

During my extensive travels, I have had the pleasure of studying breeding or wintering Hen Harriers in the neighbouring Republic of Ireland, England, and Scotland, but especially in the nearby Isle of Man, which incidentally still remains the largest winter roost of Hen Harriers in the British Isles and Ireland and probably in western Europe.

My favourite location for wintering Hen Harriers is an extensive saltmarsh in the south-west of Holland, known locally as, Het Verdronken Land Van Saeftinghe (translated as, the drowned land of Saeftinghe), a former reclaimed polder now lost to the sea. I first visited this area in February 1992 with Roger Clarke, and then on my own in February 1993. At its peak, this 2,850ha saltmarsh would host 40-60 Hen Harriers, and around 10-15 Western Marsh Harriers. Until quite recently, Hen Harriers outnumbered their much larger cousins 3/1, but due to regular intrusions by large numbers of Greylag and White-fronted Geese, which continually grazed in the harriers' roosting grounds, the trend has now been reversed in favour of the much larger Marsh Harrier. When I returned nine years later, in February 2002, I was shown this turnaround in fortunes by my good friend Henk Castelijns, but thankfully, wintering Hen Harriers continue to inhabit Saeftinghe, albeit in smaller numbers, as my last visit in 2002 confirmed.

With regard to Hen Harriers in unusual plumages, I have only observed two specimens, from 1986-2006. The first, an albinistic bird, was observed at a forest in County Fermanagh, on 2nd August 1997. There have been four previous records, but these all referred to birds in the neighbouring Republic of Ireland, one in County Tipperary and the remaining three in County Limerick. This was now the first official record for Northern Ireland.

A local forester Basil Lenaghan, had found a ground nest earlier in the season which originally contained five eggs, in a restock/clear-fell area. When four chicks eventually fledged, he discovered that one was an albino, the remaining three were in normal ringtail plumage. In flight, at least two or three secondary feathers on each wing showed dark brown shading, with small patches also noted on the crown and the nape area, at the back of its head. Overall its plumage was approximately 90% white, but it was very timid. Earlier that week my informant had watched it being attacked by a Peregrine Falcon,

A rare albino Hen Harrier   *Steve Redpath*

amazingly, it survived its first winter and returned to its natal area during the spring of 1998, but not thereafter. It is estimated, that 70% of young harriers die in their first year. Luckily this bird survived that vulnerable period in its life cycle.

My second record, involved a male Hen Harrier, in what I can only describe as melanistic plumage. On the evening of 4th July 1998, my wife Linda and I had been watching two pairs of harriers hunting close by their nest sites on a stretch of moorland in Wigtownshire, Scotland. The males, from both pairs had normal pale grey upperparts, but at 8.25pm, in good light, another extremely dark bird arrived in the area. At first glance we thought it was a sub-adult or female Hen Harrier. However, as the bird approached to 50-75m, the upperparts were seen to be of a uniform dark sooty-grey, in colour. The uppertail coverts, often incorrectly called the rump, were not white, but the same dark grey of the tail, back and upperwings. The only distinguishable features on the upperparts were the jet black primaries, with the underparts much lighter, albeit with a sooty cast.

Although there has been at least one instance of a fully melanistic young Hen Harrier, recorded at the nest (Watson 1977), this is, so far as I am aware, after extensive literature searches and consulting others, only the second record of melanism away from the nest. A Hen Harrier, which was firstly described as having black plumage with a thin pale line instead of clear white uppertail coverts, was recorded once by John Thorpe and Geoff Kelly at the UK's largest winter roost site in the Isle of Man, on 27th December 1992. But when the bird was eventually viewed at close range, its plumage was very dark grey with somewhat paler underparts and was obviously melanistic and similar in overall appearance to the Hen Harrier that my wife and I had observed in 1998. The observers were unable to judge the size and sex of the bird and no feature of the plumage gave them any indication whether it was a ringtail or adult male, but it may have been the latter. Therefore, rarity of melanism in the Hen Harrier population may suggest that it is not a viable plumage for breeding or roosting, hence the floating existence of both these birds.

Probably, one of my most amazing and unexpected records of a Hen Harrier, occurred on Saturday 13th May 1995, whilst driving in the vicinity of Titchwell Marsh, in Norfolk. Out of the blue, a female Hen Harrier, flew low across the road from the direction of the marsh and hastily disappeared over a high hedge. This was an extraordinary sighting because in May this bird should have been on its breeding grounds somewhere in northern England but not at Titchwell. I was lucky this day in Norfolk as I managed to observe all three species of harrier, that breed in the British Isles, including Marsh and the extremely rare Montagu's Harrier.

Further afield, I have observed this cosmopolitan species at Keoladeo National Park (better known as Bharatpur), in northern India, when four birds – two males and two females, roosted in the grassland, on the evening of 14th December 1997. Two over-wintering males were also recorded at Kaziranga National Park, in Assam, in November 2003, which is in the remote north-east corner of this huge country.

Hen Harriers, from below their tree nests – a personal study

Ornithology, occasionally gives us the unexpected, but surely not tree-nesting Hen Harriers in Northern Ireland. Hen Harriers have always been regarded as a ground-nesting species, so confirmation of tree-nesting at a mature conifer plantation in County Antrim in 1991 and almost certainly a year earlier in 1990, has now been hailed a first for the Western Palaearctic. This personal study, from directly below their tree nests, reveals for the first time their amazing and evolutionary struggle for survival, in what I consider, an alien habitat.

When the first ever pair of Hen Harriers was discovered tree-nesting on 14th July 1991, it was not envisaged then that numerous and unforeseen problems would surround these special nests over the following 17 years. Like so many of our conservation problems, humans must shoulder the blame for this.

If the heather moorland in County Antrim, which is this species' natural habitat had not been decimated by decades of extensive peat-cutting and over-grazing by sheep, the illegal burning of the moorland by some local farmers and widespread predation by the Red Fox, then Hen Harriers would not have been forced to take such drastic and uncharacteristic measures to survive.

If the adjacent man-made forests had not been readily available for the harriers to utilise for both ground and tree-nesting, then the population in Antrim would probably have decreased dramatically. This once superb area was always regarded as their stronghold in Northern Ireland.

As a result of this evolutionary step forward, by attempting to adapt from being a ground-nester to a tree-nester, it was inevitable that difficulties would occur for this species at their unique nest sites. The first incident occurred at one of the two 1993 nest sites. It did not come to light however until 1994 when Roger Clarke and I found a 12m high nest site. On the forest floor lay the badly decomposed, though recognisable remains of two fully-feathered chicks which had obviously fallen from a nest there in 1993. Unfortunately, this nest site had not been discovered during the 1993 breeding season.

As only one chick was thought to have fledged from the site, it was deemed necessary to regularly monitor each site from 1995 onwards. Thankfully in the event, the site eventually proved more successful with two young being observed in flight over the forest canopy, in the company of both adults.

The tree nests present between 1991 and 1994 were not monitored for fear of causing undue disturbance and possibly desertion by the incubating females and the seemingly ever present males. Only when these nests were vacated did I consider it safe to pay them a visit, when pellets, moulted feathers and even misplaced prey items were collected for analysis.

Initially I was extremely apprehensive about the new monitoring strategy we put in place in 1995, but if you remain relatively quiet and are discreet when approaching and retiring from a nest site, the adults and their broods do not suffer any form of disturbance. Visits are usually timed to take place at least two weeks after hatching. Up until this time, the chicks are fairly inactive. It is not necessary or appropriate to visit tree or ground nests during the incubation period. If the weather is exceptionally wet and windy, then all known tree nests should be visited at least once, as they are prone to collapse during inclement conditions due to their flimsy construction, resulting in chicks perishing.

A tree nest I observed in 1994 collapsed completely, smashing all five eggs. The nest site was totally devoid of strong branches and foliage, two components essential for a safe and successful nest. This was the first complete failure I had seen. It was also the first tree nest I had observed being constructed by both male and female, who diligently carried nesting material from the nearby moorland. As tree-nesting was still in its infancy in 1994 it seems reasonable to suggest that this ambitious pair was attempting arboreal nesting for the first time. Materials such as heather twigs and dried grasses that are found in ground nests, are also found in tree nests, one exception being a 23cm length of wire which I discovered in a 2m high nest in 1992!

Activity at each site during incubation is minimal, apart from the resident males arriving above the nests carrying prey for the expectant females. Persistent calling and circling low over the nest occurs to entice the female from her arboreal surroundings, with this then followed by a spectacular food-pass above the canopy or over the adjacent moorland. So agile are the females that the delivery is seldom missed and within the space of five minutes, she would return to the nest to resume her incubation duties, with the male, usually perching nearby before resuming his hunting forays. Occasionally, the male is required to drop prey into the nest, from approximately 3-4m above the canopy, when the

A female Hen Harrier contemplating tree-nesting

female refuses to leave her nest. From below, I could clearly see the male, descending in helicopter fashion, his long legs dangling down with a small prey item clasped firmly in his sharp talons. Arrivals and returns to these nests are generally carried out in this manner, as the surrounding conifers stand at least 4-5m above the actual nest site.

In 1996, a live and partially plucked Meadow Pipit was dropped into the nest by the female, but it fell through a gap in the surrounding branches and landed at my feet, some 6m below. Inevitably it died minutes later, probably of shock! At the time I firmly believed that the live and active prey item was deliberately delivered to the nest to teach the would-be fledglings (probably the eldest), the art of pouncing, and making a kill, as this important skill would be required within days of becoming independent of the nest. To my surprise, this behaviour has never been recorded at the more common ground nests, even though they have been studied by harrier experts for decades.

Then in 2002, this unique behaviour was recorded again. Only this time two live Meadow Pipits fell to the forest floor within 15 minutes of each other. Both items were delivered by the resident female. As in 1996, at least two chicks in this nest were close to fledging, hence the delivery of live prey items. Within hours of fledging, the provisioning rates at both these sites could be described as phenomenal, with both adults bombarding, the nests with avian prey. My records show, that between noon and 12.40pm at the first site, five prey items were delivered, while at the second, seven items were brought to the nest in just under an hour, 55 minutes to be exact.

From below another site in 1995, which contained three young in a perfectly built nest, Roger Clarke and I had superb views of the female mantling over her chicks to protect them from the hot sun. Thanks to the bright sunshine, the female could be clearly seen from below the 6m high nest, as she laid prone, with her body and long wings fully extended against the surrounding foliage. The summer of 1996 was exceptional, and so I had the pleasure of observing this behaviour on a further two occasions. This tree nest was the first one ever to be photographed containing Hen Harrier chicks. Mantling behaviour was also recorded, from beneath this nest. In the past only vacated nests or those containing addled eggs were ever photographed and so this one was a real bonus. Due to the excellent weather, food was plentiful; the nest at one stage containing three partially plucked juvenile Starlings.

A new strategy was implemented in 1997, with regard to prey items that had fallen to the forest floor. At two sites, they were collected and placed on mounds, tree-stumps and fenceposts by the forest edge. All the items were small birds and were amazingly utilised again by all four adults. I personally witnessed this occurring at one of the sites when observing the area from a discreet distance. In those cases where the nests were not situated close to the forest edges and nearby firebreaks (or rides), the items were completely removed and not placed elsewhere. Removing these items from the base of the tree nest, may deter foxes, stoats and other predators, as it is here the distressed chicks will remain after becoming dislodged. Although there is a case for leaving prey items below nests, as fallen chicks may be able to feed on them, I believe that the smell of rotten prey remains could attract unwelcome predators, so the safer option may be to remove them.

In comparison to monitoring ground nests, which only require a minimum 2-3 visits per season, the above clearly shows that monitoring tree nests is not only exciting, but time consuming. The reason being, that chicks in their respective tree nests, are extremely active at least two weeks prior to fledging, as they quickly gain weight and become totally independent of one another. Problems also arise, when prey is brought to the nests, which can contain up to five chicks, and when the items are hastily delivered the thriving chicks are left to fend for themselves. A free-for-all then ensues, with the eldest usually claiming the majority of the spoils and the youngest tending to be dislodged by their elder and much stronger siblings. Similar situations also occur at ground nests, but here there is usually sufficient space for quickly growing chicks to feed and fledge safely, some scuttling off into the surrounding heather to consume their prey privately. As with any ground-nesting species, they run the risk of being preyed upon by a marauding Red Fox, which are numerous in the hill farms of County Antrim. Persecution and deliberate disturbance by humans has also become prevalent here in recent years.

Subsequently, during these frenetic and dramatic periods, chicks have been observed falling from three tree nests, between 1995-2006, with similar happenings probably occurring at undiscovered sites, increasing mortality rates and decreasing fledging success. Fortunately, the majority of chicks survive, but not all tree nests are discovered due mainly to the lack of fieldwork help and the time and patience required to find tree nests. Consequently, regular monitoring of these vulnerable sites at the appropriate time, has seen 20 chicks of all ages and plumages being rescued and safely returned to the wild, which otherwise would have certainly perished on the forest floor.

Further observations from below several lower sited nests (4-8m high), revealed constant jostling for position, regular bouts of wing-flapping, brief vertical flights and chicks scurrying around calling excitedly, with the youngest uttering faint begging calls for food. The aforementioned was followed later by prolonged flights across the nest onto the surrounding branches, with fully feathered chicks observed crawling in a precarious fashion towards the edge of the outer branches that occasionally meshed with the surrounding conifers. This unique behaviour known as branching is quite common in Sparrowhawk and

A 12-metre high tree nest   *Don Scott*

Goshawk populations and although Hen Harriers belong to the same family, they have yet to perfect this trait.

It is important to add, that when chicks of all ages fall to the forest floor with the nests ranging from 2-13m high, they tend to remain at the base of the tree. From here, they can vaguely interact with the remaining chicks and possibly the adults, their sensitive hearing even picking up the feeble calls of the youngest in the brood high above them. Observations of recently fallen chicks, showed that when the adults arrived with food, the excited and begging calls of the remaining chicks are more protracted and on two occasions I have watched fallen chicks calling in response.

Contrary to what some people have said, fallen chicks will remain below the nests until they either die of hypothermia, especially young at the downy stage, with the remainder dying of starvation, or falling victim to a passing fox. After 16 years of monitoring tree-nesting behaviour, I have yet to observe fallen chicks attempting to escape to an open area (where available), where they could be fed and cared for by both adults. Basically, the survival rate for chicks below these nests is, nil! Other useful observations show that fallen chicks become disorientated and stressed, and it is likely that, due to the lack of light that reaches the forest floor, their vision becomes impaired if they remain below the canopy for any length of time. Therefore, it is imperative that any chicks found alive are retrieved immediately and appropriately rehabilitated before they are returned to the wild.

Adult Hen Harriers find it impossible to penetrate the completely closed canopy in mature conifer plantations. This is probably why attempts are never made to reach stranded chicks, even though the adults are obviously aware of their presence below. If either adult were to find themselves below the canopy, I doubt they would escape easily due to their long wings and their inability to cope with the density of foliage in these forests. Unlike the Goshawk and Sparrowhawk, Hen Harriers are a bird of the moorland and wide open spaces and are not designed for life in dense plantations. Unfortunately, drainage ditches or firebreak rides were not present alongside the majority of these nests. On one exceptional occasion, a tree nest was discovered 60 rows in from the forest edge. If chicks had fallen from this site, their demise would be inevitable. It is virtually impossible to replace fallen chicks into their tree nests because most nests are 10-13m high. Nests which are lower (2-6m) are occasionally accessible with extreme care; an adjacent conifer sometimes being utilised to view the nest and its contents. Care needs to be taken when attempting to replace fallen chicks, since it is possible to inadvertently flush those which are remaining or to damage these often fragile nests in the process.

Within hours of fledging, the chicks move to the edge of the outer branches, where a prolonged series of wing-flapping takes place, followed by extended vertical flights with several individuals rising over 2m above the nest. At one particular site, Philip McHaffie and I nervously watched two prospective fledglings crash through the surrounding branches during their kamikaze flights. On at least three occasions, the two chicks crash-landed approximately 1m below the actual nest and were extremely fortunate not to fall to the ground. After 10 minutes of branch-hopping and general unsteadiness, the two chicks eventually made it back to the centre of the nest. On the third occasion, the return to the nest was achieved rather hastily when the female arrived with a large prey item, which was immediately identified as a Snipe. The following morning having stayed in the forest overnight, the youngest of the four chicks partially in feather, was found alive below the nest.

Shortly before the act of fledging takes place, the female, will arrive above the nest calling loudly, followed by brief periods of milling (circling), directly above the site. Normally, she will be carrying prey in her talons, as she tries to coax the eldest chick to take its first flight. During this prolonged process the female remains totally focused on the chick. She occasionally dangles the prey to lure the reluctant fledgling from the relative safety of the nest, and any remaining chicks become excited at the prospect of being fed. If she receives no immediate response, she will briefly leave the area, and return minutes later repeating the same behaviour until the chick responds to her pleas. On fledging, the chick will be immediately rewarded with the prey item, which is usually a Meadow Pipit or Skylark. In one case I witnessed, the fledgling made a gallant and spectacular attempt to catch the prey at its first attempt, which quite amazingly it achieved!

With regard to the chicks that had fallen and could not be replaced into their respective tree nests, I was eventually given permission in 1997 by the DOE Wildlife Branch, to temporarily care for them at my home in Dundonald, near Belfast, prior to returning them to the wild; surely a better option than leaving them to perish below their nest. In previous years I had to travel a considerable distance to deliver dislodged chicks to hacking facilities in County Tyrone, but this new arrangement saved me a lot of valuable time and mileage. In 1995 and 1996, five chicks were liberated from these hacking pens, but sadly not into their natal home in County Antrim. I often wondered if these birds ever survived the rigours of the nearby Sperrins. Three young harriers were rehabilitated at my home in 1997 and returned to the wild a few days later. The three downy chicks of similar ages, were fostered into two adjacent ground nests, which contained a single and two chicks, respectively, of approximately the same age. To my surprise, both of these nests contained addled eggs. There were three in the first one and two in the second. Before leaving the sites, a substantial amount of food was provided for the resident females. By 20th July, all three fostered birds plus the resident chicks had fledged successfully. The five addled eggs had mysteriously disappeared, presumably eaten by the chicks, or the adults. I must add, that chicks found below tree nests, that could already fly, were fed and released immediately close to their nest site and only on the odd occasion was a chick fostered into a ground nest that already contained three or more chicks. This was the first time that the fostering of Hen Harrier chicks had been attempted in Northern Ireland and I thank my harrier colleague, Brian Etheridge, RSPB Highland, for his sound advice on the matter, and for the successful outcome. Consequently, regular monitoring of these vulnerable sites, at the appropriate time has resulted in 20 chicks of all ages being rescued and safely returned to the wild (1995-2006). They would have certainly perished on the forest floor if no action had been taken. Like their counterparts in the UK, Hen Harriers in the uplands of County Antrim and elsewhere in Northern Ireland have had many lean breeding seasons over the past 22 years. Saving two or three chicks each season may hopefully bolster the already vulnerable population.

Clear evidence of a particularly lean breeding season was first noticed by Philip McHaffie and I in 2003, when only one tree nest was found. This nest with two eggs had been abandoned shortly after incubation had commenced, due to ongoing forestry operations in the area. That year, for the first time in 13 years we could not find any active tree nests, although it is possible that the odd nest may have been overlooked. Despite the exceptional summer, no more than a handful of ground-nesting pairs were successful in County Antrim. It was concluded that the failure of the tree-nesting harriers in at least five plantations was the result of human disturbance due to car rallies and long-term forestry operations.

Persecution also blighted the season, which resulted in the deaths of three and five chicks at respective ground nests, with an incubating female mysteriously disappearing at a third site (presumably shot), and a grey male definitely shot in South Tyrone. Incidentally, forests have become increasingly important for eight of our ten species of breeding raptors and owls in Northern Ireland, as misuse of the wider countryside and particularly our uplands, has continued unabated for decades. It is well known that Hen Harriers in particular are extremely vulnerable to disturbance of any kind, and prior to beginning this book in October 2006 we have been unsuccessful in persuading the Forest Service to limit forestry operations and car rallies during the breeding season despite numerous meetings with EHS, RSPB and the Northern Ireland Raptor Study Group.

It would be unfair to bring this historic chapter to a close, if I did not briefly mention the erection of artificial platforms for Hen Harriers, another first for the species! In 1992, when tree-nesting occurred in the same South Antrim forest for almost certainly the third time, I decided, after discussion with Colin Shawyer and Roger Clarke, and later John McGhie from the Forest Service, that it would probably be advantageous for Hen Harriers, if three nesting platforms were erected in time for the 1993 breeding season. After I was granted permission from the Forest Service, my colleague Philip McHaffie and George McGrand (now ex. Forest Service) helped me on 6th Ferbruary 1993, to erect three platforms about 2m high. These were placed at various locations in the forest, but mainly in the vicinity of the previous year's nest sites.

Once these were installed, the area around the base of each platform was cleared of adjacent conifers in case chicks should fall to the ground, thus allowing the adults to feed and protect them and hopefully aid their long-term survival.

We had high hopes and expectations that these artificial nest structures would be utilised, but none have yet been occupied, despite being religiously maintained until the end of the 1996 breeding season.

As the years slipped by, I began to learn that there was no shortage of deformed Sitka's, for these ambitious birds to choose from and there may be little requirement for artificially prepared nest structures.

Top: A soaring female Hen Harrier   Bottom: Scolding an intruding Buzzard   *Sean Gray (Grayimages.co.uk)*

Since 1991, there have been many exceptional feats by tree-nesting Hen Harriers! We have recorded four instances of repeat breeding in the same tree nest for two consecutive seasons, and one record of Hen Harriers occupying a tree nest for five years (1999-2003); amazing firsts for this seemingly adaptable species. Sadly, the 1999 nest failed due to the predation of the two small chicks by a pair of Common Buzzard. Predation by other raptors is not common in Northern Ireland, but probably exists now that the Goshawk has also moved into these plantations in recent years. There is however, a record of a male Hen Harrier being found at a Peregrine Falcon plucking post by Cliff Dawson (pers comm), in the Sperrins, County Tyrone on 22nd July 1984.

Unfortunately the attempt to occupy the site for the fifth successive year in 2003, was sadly dashed at the egg-laying stage, due to tree-felling in the vicinity. At least six previously used sites were also felled at the same time.

It is quite common for ground-nesting Hen Harriers to utilise the same general area in consecutive years, so to occupy the same tree nest, for more than one breeding season is obviously a unique occurrence.

At this stage of the chapter, I must mention that all these amazing and crazy nests have been found only in the tops of deformed Sitka Spruce trees, in at least three known plantations, the furthest being over 20 miles from the original forest. When viewed from above, using an adjacent tree as a vantage point, or when standing directly below the nest, it clearly resembles an upturned umbrella without the shaft.

The last thing to evolve as a result of tree-nesting was perhaps tree-roosting by Hen Harriers. The first definite signs of this strange behaviour in vacated tree nests, were discovered by chance on 6th February 1993, when inspecting the old 1992 nest site. The ground below the 2m high nest was heavily encrusted with fresh white excreta, with several freshly dropped pellets and a secondary feather from a grey male, also present. A visit was paid to the site again on 13th February, and as I approached the nest I inadvertently flushed an adult male. This proved that tree-roosting in vacated tree nests, was now taking place. It is known that adult males rarely leave their breeding grounds during the winter months. The only exception is if the weather is bleak and food is in short supply. The males are first to be seen at their breeding grounds as they await the return of the females. To date, roosting in vacated tree nests, mainly by grey males and very rarely, by females or ringtails, has been recorded almost annually. This phenomenon has also been recorded further north in County Antrim and at Northern Ireland's only known winter roost site in the Sperrins, County Tyrone, and is probably in response to poor ground cover and potential predation by foxes.

The use of trees for both nesting and roosting gives rise to some interesting speculation on how this behaviour evolved in Hen Harriers in Northern Ireland

Which came first, tree-nesting or tree-roosting? The use of tree nests, in several of the aforementioned instances, may indicate that tree-nesting came first. However, the observations of communal roosting in trees, in both Counties Antrim and Tyrone, clearly involve much more than trees containing nests. Nevertheless, both have persisted in

Northern Ireland for at least 17 years, with hopefully no signs of either diminishing in the near future. So unique are these traits to Northern Ireland, that they have not been recorded in any of the Hen Harrier's extensive breeding and wintering ranges throughout Europe and elsewhere. Therefore, tree-nesting and tree-roosting should not be allowed to fade into oblivion by those on our island, who could not care less. The evolutionary achievements of these special birds should be regarded as exciting and phenomenal!

It must always be remembered, that during the course of each breeding season, four main factors determine whether Hen Harrier nests are successful. These are disturbance and persecution by humans, predation by Red Foxes and inclement weather, particularly during the vital incubation period. My main study area since 1986 in the magical and beautiful Antrim Plateau, has suffered directly and indirectly from all four in recent years, with other underlying problems possible in the near future. All our major conservation bodies in Northern Ireland must pull together to ensure the long-term future of this elegant and much maligned raptor, whether it is through implementing SPAs (special protection areas), or upgrading areas of land to ASSI status.

During the course of my studies I have discovered that tree-nesting by Hen Harriers is not advantageous to the long-term success and overall survival of this species in the Antrim Hills. A most recent report shows only a mean fledging rate of 1.5 chicks per tree nest compared to 2.6 chicks at ground nests (Scott & Clarke 2007). Only the return of vast tracts of deep heather moorland like those which remain in parts of Counties Tyrone and Fermanagh, will see this elegant raptor thrive in our uplands once again.

Recently, 2007 brought some hope and more importantly good news for Hen Harriers and also Merlins in County Antrim, with the implementation of the long-awaited SPAs. However the management of them has not yet been decided as EHS and the Forest Service are working together to see how they would be compliant with Conservation Regulations and the all important EU Directives. The Forest Service has also banned car rallies in forests during the Hen Harrier's breeding season (April-July), something that I have personally been asking them to do for over five years and which is only now coming to fruition. Obviously, the SPA and EU requirements have had an effect and also an immediate response from both named bodies, as failure to implement these laws could lead to severe warnings, followed by heavy fines from Europe itself. If it had not been for Europe, telling both these government-aided bodies to get their house in order (particularly the former mentioned), I believe neither would have implemented any worthwhile measures to fully protect Hen Harriers and Merlin. Hopefully, it is not too little too late for a raptor that has been showing signs of decline and sadly has not increased in significant numbers particularly in the Antrim Hills, during the past 3 to 4 years (2004-2007).

The Hen Harrier and its breeding habitat in Northern Ireland which has been neglected for decades by our main conservation organisations, is now flavour of the month here, with several researchers and numerous untrained observers attempting to duplicate the work I have been reporting on during the last 22 years. At least they should confirm the findings of my long-term research, but in the end, the future of this species depends upon how strong and committed our conservation bodies really are and whether common-sense management by the Forest Service is fully implemented.

At present I (like others), do not yet see a commitment from any of these organisations, so I reserve my judgement. In 2008 the RSPB (with others), is planning Red Kite introductions in Northern Ireland, but in my view they should make sure the future of the Hen Harrier, Merlin and Barn Owl is secure, before embarking on this project.

Hen Harriers and wind farms

Over the past four years applications for wind farm construction have become prevalent in Northern Ireland as elsewhere in the UK, with many of these pending and in some cases agreed by the statutory conservation bodies.

I consider wind farms as blots on the landscape, with the companies responsible for their overall construction seemingly hell-bent on confrontation with the general public, due to their placement in areas of outstanding natural beauty or where sensitive and vulnerable birdlife exists. In my view it is ecovandalism on a grand scale for very little long term benefit to our carbon footprint or the consumer. Wave energy is probably more efficient, less intrusive on the landscape and largely unnoticed by the public.

The statements and propaganda put out by research and wind farm companies, which suggest that these turbines are not a hazard to raptor species like the Hen Harrier and Common Buzzard etc, are in my view incorrect.

For the past two winters (2005/6 and 2006/7), Philip McHaffie and I have monitored an 11 turbine wind farm on Slievenashanaghan Hill, in the heart of the Antrim Plateau, for raptors and other birds which have succumbed to the turbine blades. The hill conveniently overlooks Slieveanorra Forest, where several species of rare raptors breed and overwinter. In the early 1990s up to two pairs of Hen Harriers bred annually on the heather clad slopes directly below the summit, but since the commissioning of the wind farm they have left the area in favour of the nearby forest.

Since October 2005, we have found on our visits, the remains of eight badly mutilated Hooded Crows and six Ravens, which had collided with the turbines, but never a bird of prey. On the morning of 13th January 2007, however, our visit revealed the desiccated remains of an adult male Hen Harrier below one of the turbines, which was not apparent during our last visit on 10th December 2006. Thankfully, the long dead harrier had not been scavenged by foxes, allowing us to inspect the carcass, which was still recognisable but had deteriorated badly. The right wing had been completely sheared off by the impact and lay approximately 4m from the body.

Around 11.00am on the morning of 16th February 2006, we arrived in the same area to carry out our monthly check for recent bird kills. On arrival, we were somewhat surprised to find a large white van parked close to one of the turbines and were just in time to see him deposit, what was almost certainly a dead Raven in the back of his unmarked

Female Hen Harrier at the nest

van. At the time we thought he was a maintenance engineer, as he was wearing navy blue overalls and he appeared embarrassed by our sudden appearance, which seemingly forced him to leave the area promptly.

Could he have been from the wind farm company responsible for the 11 turbines in this area and was one of his duties the collection of casualties that had collided with the huge blades? Possibly yes, as we seem to hear these companies declare that wind farms are safe for birds, so if none are found then they must be safe. I mentioned this particular incident to several birding friends and also to RSPB-NI, who apparently rang the company in question to enquire were they responsible for removing dead birds from below their turbines? The answer was an emphatic no, but then are they really going to admit that they do?

Long term studies of harriers in this area (Scott & McHaffie), have consistently shown that since the erection of these turbines they purposely avoid them when entering or leaving the confines of the adjacent forest or when foraging nearby, so it is likely an over-wintering bird unfamiliar with its surroundings will accidentally fly into one of the turbines blades in dense fog. During a follow-up visit on 1st February, the weather that day was fine and sunny with no wind, but all 11 turbines were completely shrouded in mist, so this area can be deceiving for numerous bird species even in favourable conditions.

Elsewhere, surveys over several decades at wind farm sites in North America, continental Europe and most recently Japan, report that only large raptor species such as, eagles, vultures and kites, were being killed in large numbers, but not apparently smaller falcon and hawk species. In Scotland, studies of Golden Eagles in Argyll by Mike Gregory (pers comm), tell us that these magnificent raptors have either deserted their breeding grounds or failed to breed in recent years due to the arrival of wind farms. The furore caused by the proposal to erect 181 turbines on the Isle of Lewis, should be a warning to everyone who is interested in conserving our wildlife and in particular our rare bird species.

Quite simply, wind farm companies now need to go back to basics before declaring that these structures are safe for raptors like Hen Harriers. Research work investigating the potential impact of wind turbines on birds is in its infancy in Ireland, therefore, further comprehensive and detailed studies need to be carried out forthwith before licences are granted to construction companies whose sole interest is profit.

It is not good enough to say that Hen Harriers tend to fly at low altitude and appear to be less vulnerable to collision than much larger raptors. There are also other implications that must be considered such as: displacement, considerable loss of habitat, disturbance, prevailing weather conditions and why crow species appear to be most vulnerable to collision. When all these questions are satisfactorily answered, then we can perhaps consider whether wind farms are what we really want on our wild and desolate uplands. In recent years, I have been surveying proposed wind farm sites in both the northern and southern counties of Northern Ireland for breeding Hen Harriers and other high-risk species. This I have done impartially and with an open mind, despite my long-standing abhorrence of wind farms.

No doubt applications will be considered in the near future and then granted by those in authority to do so, but hopefully the above conditions will be fully addressed otherwise we may encounter problems that will be detrimental to rare breeding species in our uplands. With the vital SPAs now in place, wind farm companies will probably be forced to utilise peripheral sites that are away from the main core breeding areas for Hen Harriers and other related species, and rightly so. Most recently, at the behest of RSPB NI, I twice registered my objections against the proposed wind farm scheme on the Isle of Lewis – so if their UK consultants consider them a danger to birds on that island, which hosts several highly important breeding species, then why are we not taking the time to consider them here in Northern Ireland? Are turbines not a danger to birds here as well? Surely the long term future of our wildlife is more important than wind farms, is it not?

Hen Harriers: polygyny versus polyandry – who wins?

Polygyny (where one male breeds with several females), is rarely observed in the Northern Ireland Hen Harrier population. Eddie Balfour before his untimely death in 1974, regularly recorded this characteristic feature of their breeding ecology on Orkney. This was also recorded by Nic Picozzi between 1975 and 1981. In all these cases there was an unbalanced sex ratio, with approximately 2-3 females to every male, and so on Orkney it was not uncommon for a single male to have three, four and even six females, during the course of one breeding season. Interestingly, their nests were extremely close together in an area that would normally be occupied by 1-2 pairs, which finally resulted in a high incidence of breeding failure. Thankfully, on Orkney today there is now a normal Hen Harrier population.

On the other hand, polyandry (where one female breeds with several males), has until recently never been recorded in the vast amount of Hen Harrier literature. However, females participating in this, the more uncommon form of the two behaviours, have been recorded in the Antrim Hills population, but their breeding attempts, particularly at tree nest sites are usually problematic and rarely successful. Therefore, my knowledge of both is limited and why it is now occurring with regularity is baffling to say the least. As a consequence, very little information exists regarding the frequency with which this now occurs throughout our island, especially in the uplands where it probably goes unnoticed, due to the lack of experienced observers.

This strange behaviour was first recorded here at Ballypatrick Forest in North Antrim by Fred Quinn from Forest Service (pers comm) in 1978, when a polyandrous adult female had three grey males attending her at a ground nest. At one stage, the nest contained five downy chicks, of which two unfortunately died, with the remainder fledging successfully.

In 1995, I recorded my first ever polygynous Hen Harriers at a mature conifer plantation in south Antrim, when a male whose own ground-nesting attempt had failed, joined forces with a neighbouring tree-nesting pair for over six weeks and then disappeared shortly after the three chicks had hatched. All three chicks in this instance fledged successfully, thanks to the vast amounts of prey brought to the nest by both males and later by the resident female.

At the same forest in 2004, one pair had tree-nested by 8th May, with the remaining two males and a polyandrous female confirmed arboreal nesting on 9th. The latter had occupied a tree nest that was last utilised successfully by what I regarded then as a normal tree-nesting pair, during both the 1995 and 1996 breeding seasons. By 26th May, no activity was recorded at the site or on subsequent visits in early June and so I visited the nest to ascertain why it was now deserted. On the forest floor lay two freshly broken eggs, with a scarcity of nesting material on the deformed trunk and surrounding branches. It appeared that the eggs had been laid in haste by probably a female with no experience of tree-nesting, and with both males also devoid of arboreal skills, their random efforts to occupy this site were abysmal. Little did I know at the time that I would observe a similar situation to this during the 2006 breeding season.

As the previous attempt was doomed to failure from the start, I was then informed by Fred Quinn, on 1st July 2004, of another polyandrous female, who also had two males accompanying her at a ground nest in the same North Antrim forest as that of 1978. The nest again contained five chicks, and by 18th July all five were observed on the wing in the company of both adults. For the record, this forest was over 20 miles from the previous location!

Regular visits to this forest by Philip McHaffie and I earlier in the season, appeared to show that two pairs were consistently holding territory, but we were totally unaware of this second polygamous breeding attempt. In addition, our pre-breeding season recces in search of territorial Hen Harriers in County Antrim, did show an excess of males, but this tends to be a normal occurrence between late March and mid April, as males are known to return to their breeding grounds much earlier than females.

Elsewhere in 2004, a polyandrous female was discovered in the Forest of Bowland in Lancashire, which was probably the first record of this behaviour in the UK. During the 2002 and 2003 breeding seasons, circumstantial evidence suggested that a female in this forest was polyandrous, but as the 2004 season progressed the two observers (Bill Hesketh and Bill Murphy) now had proof that she had mated with more than one male. After eight hours of observations on 23rd April, they recorded that she had copulated and took prey from three grey males, two of which were wing-tagged. The nest contained six eggs when first checked on 26th April, but sadly their report does not tell us if the nest was successful. Like me, the two Bill's, did not know of any previous published records of polyandry, and wondered if it was caused by an unbalanced sex ratio or the two non-dominant males being unmated and simply attaching themselves to an established pair.

After the unforeseen events of 2004, I expected 2005 to be a normal year for Hen Harriers in the Antrim Plateau, but then how often since 1990 have these birds ever conformed to a normal breeding season in this county – never! As the season progressed it soon became evident that the two known traits (polygyny and polyandry) were occurring simultaneously in two south Antrim forests which were no more than 5km apart, as the crow flies.

Once again five Hen Harriers, not four, were attempting to tree nest in the same forest, as in 2004. By 11th May, one pair appeared to be safely ensconced in a tree nest that was last occupied in 2002, but by 5th June the nest had failed. The polyandrous female and the two males at the second site had also abandoned their nest by 12th June, with both failures due to a combination of harassment at the former by a pair of Common Buzzard and wet and cold weather for much of May to mid June. The sky-dancing displays performed by the two males when the polyandrous female appeared were unbelievable! A visit was paid to both nests which were still perfectly formed but no broken eggs were discovered at either.

Nearby, at a relatively young conifer plantation the more common form of polygyny was unexpectedly occurring. In early May, one pair had just commenced ground nesting at the south end of the forest, but on 15th May I discovered a second female was also ground nesting 500m further north, with no sign of an additional male. Further observations clearly showed that this male was provisioning both these females. By 21st June, the polygynous male had lost his second partner, but was still providing food for the original female who by now had three chicks. Once again the presence of two Common Buzzards, may have contributed to the demise of this second ground nest, which contained two eggs on 25th May, as the lone male was never observed defending the site when the female was absent. The original nest was of course successful with all three chicks, two males and a female already fledged by 11th July. The discovery of these two forms of polygyny, occurring within 5km of each other is probably unprecedented for this species in the Western Palaearctic!

The 2006 breeding season was even more intriguing, despite the poor weather that badly affected the vital incubation period, resulting in several complete failures, small broods, numerous addled eggs, but more importantly poor fledging success.

Two pairs were again present in the same south Antrim forest during April and early May, where the act of polygyny, has now become something of an annual event. By 11th May, one pair had seemingly ground-nested at the south-western corner of the forest, but at the northern end two grey males were relentlessly pursuing the resident female across the forest and the adjacent moorland. On 16th May, the brace of males, one of which was a pale plumaged old bird, and the other a darker and much younger individual, were observed circling over a block of tall conifers beside the forest path, but there was no sign of the female! Could this be another, or perhaps the same polyandrous female, that tried unsuccessfully to breed here in 2005?

Some 20 minutes elapsed and then the female ascended from the trees calling loudly in an agitated fashion. When she failed to return to what may have been a tree nest, Tim Hipkiss and I agreed that we should try to locate the potential nest site. Thirty minutes later, we discovered a freshly broken Hen Harrier egg lying on the forest floor approximately 100m in from the forest edge. As we peered up at the 10m high Sitka Spruce, we could find no signs of a nest but we did discover an array of side branches which clearly resembled one and it was below this tree that the egg was found.

When we returned to the same area on 22nd May, all three birds were continually milling over the trees on the opposite side of the path and probably no more than 200m from the first failed site. From a prominent vantage point further along the forest path, the female suddenly disappeared down into the trees where she was possibly tree-nesting again. The two males flew off in different directions, but within minutes she was in the air again and similar to our observations on 16th May, she did not return to the new potential nest site. After 15 minutes, with still no sign of the female, we again searched the area and 20 minutes later we discovered another two broken eggs lying below an 11m high Sitka. Once again, no nest was found, but the formation of the side branches was similar to those found on the first tree six days earlier.

In early June, the female was observed again, but on this occasion only one male, the paler and more mature bird was attending her, as the dingy plumaged younger male had unexpectedly moved to a ground nest site less than 1km away. The pair had now moved to an adjacent block of conifers less than 100m from the previous (second) site and only 300m from the original location. After several sightings of both birds, including four spectacular food-passes, up to and including 18th June, we decided it was time to launch a search for the nest after another airborne delivery on 23rd June. The nest was found after a short search and was only 10 tree rows away from a nearby drainage ditch and 38 rows in from the forest edge. The nest was 11m high and was precariously placed on the bole and side branches of a badly deformed Sitka Spruce. Poor tree selection and flimsily constructed nests have previously been problematic in County Antrim and so this particular site did not appear too promising, considering this female's poor choices so far.

In brief, the nest began to disintegrate due to heavy rain and on 17th July, a chick in down (7-10 days old), was found dead below the nest. By 27th July, the nest had split in two and on the 30th a fully-feathered chick which could fly, was found in a drain a short distance from the nest. On 5th August, the last of the three chicks was also found on the forest floor. This chick was also capable of sustained flight, but was impeded in doing so by the low branches of the surrounding conifers. Both were females and extremely aggressive, but they were ably provisioned, prior to release, with discarded prey items, namely, Meadow Pipit, which had also fallen out of the collapsed nest. On 8th August, the two rescued fledglings were observed within metres of each other, after being fed by both adults during a double food-pass. The two earlier failed attempts were similar in every way to what I had witnessed in 2004; perhaps the same trio of birds were involved in that disaster as well?

At present I can only speculate as to why polyandry is now consistently occurring in the Antrim Hen Harrier population (and to a lesser extent polygyny), but recent evidence suggests there was probably a dearth of females in this area between 2004 and 2006,

Top: Slieveanorra, Northern Ireland – tree-nesting habitat   Bottom:Hen Harrier chicks in a tree nest   *Don Scott*

which had gone unnoticed. Hopefully, further studies of this behaviour will reveal the real reasons for its occurrence and what implications it might have in the near future. If we are to believe the high, and in my view, the inflated numbers produced by the RSPB during the 2004 UK Hen Harrier Survey, then harriers in Antrim do not have a problem! A similar breeding survey was carried out simultaneously by Philip McHaffie and I who have studied Hen Harriers in the Antrim Hills for 22 and 19 years respectively, indicated that the RSPB's numbers were overestimated by 30% in this area alone. With regard to polygyny and polyandry, there are no winners where this behaviour persists, in what is usually a stable and unique population!

The following was written in 1995 by John McGhie before his untimely and sudden death at only 47 years of age. It was intended for inclusion in this book. John, was not only a good and obliging friend, he was a visionary during his time with the Forest Service in Northern Ireland, but sadly there are too few of his ilk left in this organisation today. John resigned from his post as Wildlife and Conservation Officer, in December 2000 and immediately commenced work in January 2001, as the RSPB's Area Manager for Islay, Colonsay and Oransay. With kind permission from his wife Susan, I personally dedicate this chapter to him in appreciation of his unstinting help with the tree-nesting Hen Harriers in the forests of County Antrim and for conserving a Hen Harrier winter roost site, the only one currently known in Northern Ireland within a forest, high in the Sperrins, County Tyrone.

Forest management and the Hen Harrier
Background

The relationship between the Hen Harrier and forestry can be traced back to the years immediately following the end of the Second World War. At that time, afforestation underwent rapid expansion in the UK, mainly in potential Hen Harrier habitat in the uplands.

The driving force behind this large-scale upland tree planting, was post-war government policy which had the dual aim of achieving self-reliance in both timber and food in the event of another war. The heavy losses suffered by the Atlantic convoys transporting supplies into the UK during the war had the effect of making the immediate post-war government determined at all costs to avoid being dependent on imports in the future. However, agricultural land was reserved for food production and so the expansion of forestry was achieved in the less productive uplands.

Forestry planting in the post-war years was conifer dominated. In the majority of upland forests a Sitka Spruce monoculture prevailed, broken only by the presence of some Lodgepole Pine. These new, and often large, forests were planted on the expanses of moorland which provided both nesting and hunting areas for harriers.

Upland conifer forests are slow to establish, and the time from planting to the clearing of one crop can vary from 40-60 years in many forests. The immediate impact on Hen Harriers, therefore, would have come not from the presence of the small newly-planted

HARRIERS: JOURNEYS AROUND THE WORLD

trees, but from potential disturbance caused by the appearance on the site of foresters and their machinery. Preparation of the ground and the planting of trees is a seasonal activity, lasting only a few months and normally ending in March or April, coinciding with the return of Hen Harriers, to their nesting grounds. It is likely therefore that the establishment of these forests had very little long-term negative impact on Hen Harrier populations. In fact, the change of land use in some areas from grouse moor to forestry may have helped to reduce persecution of the birds by gamekeepers and improved harrier ground-nesting habitat by removal of over-grazing. The new forests were relatively quiet places and the chances of disturbance during the nesting season were greatly reduced. Food supply was also likely to be more easily obtained as small mammal and passerine populations increased in numbers. This combination of circumstances should in theory have lead to greater breeding success. The utility of young forest to ground-nesting harriers would have lasted until the trees closed canopy and the harriers could no longer easily reach the ground, normally 10-15 years after planting.

Large-scale upland afforestation virtually ceased in the 1980s due to the abandonment of post-war policy, the availability of money to acquire better agricultural land and environmental pressure to preserve the remaining heather uplands. With the exception of parts of the Republic of Ireland, most current afforestation is taking place outside the Hen Harrier's upland breeding areas.

The early post-war upland forests are now reaching maturity. Until the trees are cleared and the ground is replanted these forests have traditionally been empty of harriers for some 30-40 years, as they are generally unsuitable for nesting and hunting. Hen Harriers which nest in recently cleared forests appear to do so as successfully as those which nested in the forests when they were newly planted.

Forest management considerations

With the loss of so many of their traditional upland breeding sites to forestry and grazing, those harriers which have taken to nesting in plantations make up an important part of the breeding population and the status of the population as a whole may well be dependent on the breeding success within forests. So what can foresters do to protect these breeding birds?

The most important aspect is to be aware of their presence and the timing of forestry operations can then be adjusted accordingly.

Winter roosting sites within forests, both lowland and upland, are normally occupied between September and March. There are two important aspects to managing these sites. Firstly, they are normally open areas and may require to be maintained in that condition through the removal of natural conifer seedlings and broadleaved scrub which tends to develop after the cessation of grazing. Additionally, the sites should be kept free from disturbance, which assumes greater significance when sporting rights are let and there is a risk of the birds being disturbed by dogs and guns. Winter roost sites should be designated as off-limits in such cases. If tree-roosting should become established, harvesting operations in these areas would require to be re-scheduled in order to avoid unnecessary disturbance.

At ground-nesting sites, planting of trees is normally finished by the time nest-building begins, usually in late April or early May, although there may be some overlap with courtship displays. Incubation often begins around mid May, lasting some 30 days. This is a key period as it coincides with the ideal time for herbicide application around newly planted trees. Chicks have normally fledged by mid July and are reasonably independent by August. An application of herbicide at this time would still be effective and should be used as an alternative to earlier application where nests are known to be present.

At tree-nesting sites, the trees selected by the harriers for nesting vary in age and height, but so far have two features in common. All are Sitka Spruce and all have suffered damage resulting in loss or diversion of the leading shoot. This has the effect of creating a natural hollow in an otherwise uniform tree canopy when viewed from the air. The implication is that harriers view the solid mass of trees below as the ground (which they are unable to see). Theoretically, therefore, harrier tree nests could be found in any age of conifers after canopy closure, ie, from 15-20 years of age onwards, depending on the rate of tree growth.

In Northern Ireland, our early experience of tree-nesting by Hen Harriers has already highlighted a number of forestry activities which could pose a threat to successful nesting. Certain operations have been moved to different times of the year in order to avoid disturbance to the tree-nesting Harriers. Application of fertiliser by helicopter was brought forward to late March/early April. Interestingly enough, the harriers flying over the forest completely ignored the helicopter, even when it flew within 300 m of them. This however is not recommended at sensitive stages in the breeding cycle! Other operations which have been re-scheduled include the cleaning out of drains and the cutting of access routes (inspection racks), through the forest. The most obvious activity, which has not yet had to be carried out close to tree nests, is the measuring and felling of the trees.

The preparation of the ground for re-planting can be more of a problem for nesting harriers, than it was first time around. The preparation of the ground itself can take longer, as branches and tree stumps have to be dealt with as well as the usual drainage and weed problems. This combination of circumstances makes an overlap with the harrier's breeding season more likely on these sites.

To summarise, providing that foresters are aware of the presence of Hen Harriers, in their forests, affording them a degree of protection is not an onerous or expensive task, as all normal forestry activities can be carried out at some stage of the year.

John McGhie: August 29th 1957 – March 18th 2005

Female Western Marsh Harrier alighting at her nest

3

WESTERN MARSH HARRIER AND ITS SUBSPECIES

Circus aeruginosus and Circus aeruginosus harterti

The **Western Marsh Harrier** is one of the most widely distributed of 'Old World' raptors. In the UK and western Europe it is found breeding in lowland and open country where wetlands, either fresh or brackish exist, but particularly in coastal reedbeds. Its breeding range extends to almost every country in Europe and across the Palaearctic eastward to Lake Baikal, where it is eventually replaced by the Eastern Marsh Harrier.

In Ireland, the Marsh Harrier was formerly a widespread and abundant species, but well before the end of 19th century it had ceased to breed in Northern Ireland and in the Republic of Ireland by around 1917, with single birds seen up to 1935. Elsewhere in the UK the Marsh Harrier population had slumped to a single pair by 1971 and appeared to be heading for extinction for the second time. Remarkably, by 1995, this once vulnerable UK population had recovered to 156 breeding females. In 2005 this figure had swollen to an estimated 360 breeding females, in England and Scotland, a 131% increase from the previous count a decade earlier. This was the largest number of breeding Marsh Harriers, recorded in the UK for at least 200 years, with a minimum of 800 plus chicks fledging successfully.

This record breeding success over the past 34 years has been phenomenal and although traditional reedbed sites, most of which are protected areas, continue to be regularly utilised, Marsh Harriers are also nesting in unprotected farmland amongst arable crops. The overall success of this species in the UK may, subject to the conservation of existing reedbeds and the creation of further suitable habitat, further enhance the possibility of breeding in Ireland, north and south. With good numbers of mainly female and immature birds now occurring each spring, all it will take is a single male to remain on site and then we will be able to add the Marsh Harrier to the Irish breeding list once again.

My first ever sighting of a Marsh Harrier in Northern Ireland was not until 16th May 1995, when a female/immature bird was seen in unfamiliar habitat; it was slowly quartering over the moorland in the uplands of County Antrim, in breeding Hen Harrier territory! This was quickly followed by a second bird, which frequented the rough ground close to the George Best Belfast City Airport in County Down on 27th April 1996.

Prior to these two important sightings, I had previously observed this species at Saeftinghe, a coastal saltmarsh in southwest Holland, in February 1992, when only 15-20 birds roosted there, but independent from the wintering Hen Harrier population which numbered around 40-60 birds at that particular time. I returned to the same area in February 1993 and found that Marsh Harrier numbers were already showing a slight increase since my visit the previous year. There were now around 30-40 birds present, with the spectacle of both species roosting here each evening, a joy to behold for any visiting birder. In February 2002, during my third visit to Saeftinghe, Marsh Harriers now dominated this superb area, not only in the evenings, when 50-60 roosting birds were a common sight and up to 80 on occasions, but also during the day as most birds preferred to hunt within the confines of this large saltmarsh. Almost all were females

Male Western Marsh Harrier near the nest

Niels De Schipper

and immature birds, but irrespective of this, what an impressive sight on cold and blustery February evenings as they congregated in one particular area of the marsh, before finally roosting in the tall Phragmytes.

My next exciting encounters, with this the largest of Europe's four species of harrier, were at the Parc Natural De S'Albufera, on the island of Mallorca in late May and early June 1994. The Parc is an important site for both wintering and breeding Marsh Harriers and when courtship displays commence in early February, around 10-12 pairs are usually present each year. Nests are built by the end of March, sometimes earlier, and by the time I arrived for a family holiday, chicks had already fledged from several nests with others being fed continuously by the ever present females. The majority of males of this species appear to leave the area soon after the young have hatched leaving the females to rear them. Out of the 13-16 adults present that I observed in the Parc at this time, only 3-4 were males. One particular male seen carrying food to a nest was obviously from a late breeding pair. I have also noted this behaviour by the occasional male Hen Harrier at both ground and tree nest sites in County Antrim, in recent years.

The harriers were active from shortly after sunrise and seemed to start hunting almost immediately. Individual females, recognisable from plumage characteristics, frequented the same areas of the marsh throughout the day. By mid morning hunting activity ceased, but several females would then take to the air and commence circling on the warm thermals before dropping down unannounced into the tall reeds. Other flights during the early afternoon tended to be no more than a gentle glide or a brief circle over the reedbed.

Activity would gain momentum again later in the afternoon and continue through the evening until just before sunset. I found that a good time to watch these birds hunting and then settling into the marsh was between 6.00 and 8.30pm. Seats located along a new road a short distance from the main entrance provided me with excellent views of an adjacent canal, a lagoon and a large section of the marsh. Foraging harriers flew about 2m above the reeds, with no more than three wing-beats followed by a short glide. They methodically quartered the area, often returning to places they had already passed over, stalling momentarily in the air if they located prey. When prey was about to be seized they dropped quickly into the reeds, their wings firmly held upwards, their long legs thrust downward, with the head and eyes completely focused on their quarry. They often remained in the reedbed for some time, partially feeding themselves before carrying the remainder back to their respective nests. On landing at the nest, they would hover for a few seconds in helicopter fashion with both legs dangling freely, irrespective of whether they were carrying prey and occasionally they would perform an unrehearsed 180 degree turn.

On the evening of 1st June, I watched a female leave the marsh and begin quartering along the edge of the large lagoon which held good numbers of Mallard, Coot, Moorhen and 50 plus roosting gulls. The gulls immediately scattered in all directions as she launched an unprovoked attack across the shallow water, her long legs fully extended once more and as the much slower Coot tried to disperse she cleanly lifted an almost fully grown chick out of the water. By committing this act, she was forced to beat a hasty retreat into the nearby reedbed, surrounded by large numbers of Black-headed and Herring Gulls.

These large raptors appeared to have very few natural enemies, but Black-winged Stilts regularly harassed them and the occasional attack from the previously mentioned gull colony never amounted to anything spectacular. In fact these birds always appeared wary of any bird that unwittingly crossed their path. On the very odd occasion when gulls engaged them, the harriers would simply weave their way out of danger without even beating their powerful wings. For their size they are still elegant and superb fliers and their turn of flight is deceiving and appears lethargic to those observing this species for perhaps the first time. Certainly, their dominating presence at the S'Albufera was one of the main highlights of my holiday.

One point worth mentioning is that they seldom left the confines of this huge marsh to hunt elsewhere. An exception to this occurred at 7.00am on 4th June, when a lone female was observed leaving the marsh to forage over a large area of dry scrubland adjacent to the Parc and close to the power station on the S'Albufera Pobla Road. The bird hunted in this area for 15 minutes before returning to the marsh, alas without prey. She hunted low over the ground in a manner similar to that of Hen Harriers, proving the adaptability of this species when required. These were the true denizens of the S'Albufera Marsh, despite the presence of Osprey, Booted Eagle and several rare Eleonora's Falcons.

On 12th and 13th May 1995, when visiting Roger Clarke at his home in Cambridgeshire, we spent two glorious days observing breeding Marsh Harriers in this county and also in nearby Norfolk and Suffolk. Near Ely, I was amazed to see two pairs nesting contentedly side by side in a disused sugar beet pond which was probably no larger than a small size swimming pool. Both males were particularly active and fresh nesting material in the form of small sticks and dried grass were brought to what could now be described as a reedbed.

At Titchwell Marsh, this species was highly conspicuous, with both males and females patrolling the vast reedbeds on a regular basis, with the former observed passing food to their respective partners on several occasions before the latter disappeared into the extremely tall Phragmytes. We spent over four hours at this location just sitting watching the comings and goings which were mesmerising to say the least. Around the Suffolk area the sightings were of a similar nature and probably more personal than those I witnessed a year earlier in Mallorca. At all the previously mentioned locations, you really could get reasonably close to these magnificent birds without intruding into their territory and causing them undue disturbance. They were two of my most memorable harrier days!

The only Marsh Harriers I have ever observed in Scotland were a breeding pair at Insh Marshes in Speyside, during a week long break to the Highlands in late August 1996. This was the first proven successful breeding for the Highland Region, with two fledged young. The male arrived on 24th April, followed by two females on 2nd May, with one of the females leaving the area a few days later, the rest is pure history. I

Male Western Marsh Harrier, nest-building

Ludo Goosens

watched these birds intensively over a two-day period and was privy to observing both chicks being fed on the wing by the resident female. The male had already left the area two or three weeks earlier, which is common shortly after fledging with most harrier species, so I was not too disappointed by his absence. Although the occasional Marsh Harrier has summered at Insh Marshes since 1966, it was 30 years before a pair actually bred, which was probably a relief for everyone involved at the Reserve. Disappointingly, in 1997, there was no repeat performance even though a female was present from 20th to 26th May, but sadly no male.

During my first visit to India in December 1997, wintering Western Marsh Harriers were a common sight at Velavadar National Park, in Gujarat State. The large grassland roost site was shared with Montagu's, Pallid and the occasional Hen Harrier, with the much larger Western Marsh, making up about 5% (150) of the 3,000 plus birds that were present each evening. During the day most males and females could be seen frequenting the numerous wetland areas adjoining the park, where waders and wildfowl would be hunted by these impressive raptors. Also at the world famous Bharatpur, in Rajasthan State, Marsh Harriers were much in evidence during the day at this superb wetland. Birds could be observed perching in bushes and even in the branches of tall trees that overlooked the various lagoons which were teaming with wildfowl during the winter months. At the grassland roost within the park, there were over 30 Marsh Harriers present on the evening of 14th December, which included six males and several juveniles with distinctive rufous head markings; the remainder being adult females.

When I returned to Velavadar in late November/early December 1999, and similarly in 2000, the whole area was suffering from a severe drought. Consequently, there were very few wintering Marsh Harriers present in 1999. In fact, the numbers of harriers present in 2000 had decreased from 1,100 birds in 1999 to just over 200, of which none were identified as Western Marsh! During my extensive travels in both years throughout the State of Gujarat, only small numbers were recorded as the severe drought continued unabated, which was most disappointing.

From 1999-2006, I have visited the Gambia in West Africa five times for a family holiday. This small country bordered by Senegal remains one of our favourite holiday destinations, and although it is renowned for its excellent birding we met several lovely people there as well. One in particular, Mass Cham was not only our birding guide he was also a dear friend of the family and sadly he passed away shortly after our 2006 visit.

During each holiday, Mass would take me to the 'Raptor Track', spending many hours in the field concentrating on birds of prey and on most visits we would always encounter a Marsh Harrier or two. They were always females or immature birds and they were never foraging near marshes or wetland areas. All these birds were observed at inland locations, hunting over long grass, scrubland, small bushes and also semi-desert areas, which again showed their adaptability to the conditions that prevailed.

A favourite haunt for Marsh Harriers many years ago was the Hula Swamp and Lake

Sub-adult male Western Marsh Harrier

W. S. Clark

in Northern Israel. Then in 1953/54 the marsh and part of the adjacent lake were supposedly reclaimed for agricultural production and also to rid the area of malaria, with the 10-15 pairs of Marsh Harriers that bred there apparently poisoned. These birds were exceptional in that males and females had similar plumage details and it is possible that they were members of a separate and unique subspecies which were cruelly exterminated before it could be properly identified. A similar phenomenon is believed to have existed among Marsh Harriers breeding in the swamps of Iraq, but these birds have also dispersed or been extirpated. The area contained today (some 400ha) in the Hula Nature Reserve, Israel, is all that is left of the once extensive marshes that completely filled the Hula Valley; needless to say Marsh Harrier numbers have declined as well. I would love to have had the chance to observe these unique members of the genus *Circus,* before their demise!

Dark-phase Western Marsh Harrier   *W. S. Clark*

Although it is always nice to see the Western Marsh Harrier in other countries, it would also be exciting to see it being restored as a permanent breeding resident in Northern and the Republic of Ireland, once again. I yearn for its return, which hopefully will be sooner rather than later!

The race ***Harterti*** is a subspecies of the Western Marsh Harrier. It is a resident in northwest Africa, mainly in Algeria, Morocco and Tunisia. It is sparingly found breeding in the coastal marshes close to the Mediterranean and also at altitude by lakes in the Moyen Atlas region at up to 1600-1800m above sea level, but also in drier areas with

♀ & ♂ Marsh Harriers ss Harterti
Morocco   P.Snow.

Subspecies of the Western Marsh Harrier – Harterti   *Philip Snow*

suitable habitat. As this chapter shows, small numbers occasionally migrate northward annually, with one particular bird found breeding on the Balearic Island of Mallorca.

When on holiday in Alcudia, Mallorca, from 29th May-10th June 1994, our family apartment conveniently overlooked a large section of the "Parc Natural De S'Albufera," which is better known to British and Irish birdwatchers as the Albufera Marsh. At such a southerly latitude there were no breeding Hen Harriers, but the S'Albufera, is an important site for breeding Western or European Marsh Harriers, with at least ten breeding pairs.

During the first week of my holiday, I regularly observed a female, with distinctly different plumage markings to that of the other resident female Marsh Harriers. She had a very white head and throat and extensive creamy-white patches on both upperwings and also her back and breast. The remainder of her upperparts were chocolate brown, with her belly and underparts much lighter and rufous when viewed at close range. My best views of her were from the balcony of our apartment as she hunted the extensive reedbeds directly behind the tennis courts. This location was regularly patrolled each morning and then later in the afternoon, as she had apparently nested nearby.

As the week progressed, I had made friends with several of the Parc staff and in particular Pere Vicens, who is responsible for Marsh Harrier conservation. Pere explained that the unusual Marsh Harrier I had keenly observed every day, was a female of the North African race-*Harterti*! The occurrence of this race, which is more common in Algeria, Morocco and Tunisia, was regarded by Pere as a very rare event for the Parc! Pere then went on to explain that this female had mated with a second-year male of the nominate race *aeruginosus*, and it was from her nest that the first juvenile of the year had fledged, on 8th May. This was a beautifully marked bird and its plumage was in total contrast to females of the common nominate race, which I have observed elsewhere many times.

This bird had been present at the Parc for four years and was clearly visible and recognisable to those who are seriously interested in the genus, *Circus*. At this stage I must add, that the Western or European Marsh Harrier, has one recognised subspecies – *Circus aeruginosus harterti*, which as I have already mentioned is resident in northwest Africa, so for this female to return to breed at the S'Albufera for the fourth consecutive year, is a major coup for the Parc and its staff.

In July 1995, during a two-week family holiday to Port El Kantoui, in Tunisia, North Africa, I reluctantly hired a car for three days (which was very expensive, even in 1995), and went in search of this extremely rare and little known harrier species. After driving for seemingly miles and miles, my wife Linda and I arrived at this lush green wetland well away from the main road, in an unknown location, as we did not have the benefit of a map. Within minutes of our arrival, we were immediately rewarded with excellent, but somewhat distant views of firstly two males and then a similar number of females, of this subspecies, as they quartered the area in search of prey.

The males had much darker upperparts and a slightly paler head and seemingly lighter underparts than the nominate race, with the female's plumage as described earlier in this chapter. When the male and female were seen close together and compared with the same genders of the Western Marsh Harrier, both appeared noticeably smaller. There was at least two pairs present at this unknown site, which we visited again the following day, with our observations on this occasion more distant, but just as important. As it was early July, I expected to see recently fledged juveniles, but sadly none were encountered although they may have been present amongst the tall reeds and greenery. Other areas with suitable habitat were visited on the third day, but sadly no harriers of this race, were observed.

Perhaps this was a true reflection of just how rare and vulnerable this subspecies is in this part of Tunisia and probably elsewhere in northwest Africa. This is probably due to the loss of their marshland habitat for foraging and nesting, with tourism now increasing and new hotels being built in several notable areas where harriers and other wetland species possibly bred in the past. Evidence of this was observed on the first day, as two of the birds continually foraged over adjacent dry areas containing long grass and scrub, in search of small birds and most certainly mammals. Therefore, could the loss of suitable habitat, be the real reason why this female, moved permanently to the, S'Albufera Marsh, in Mallorca? There is very little relevant information about the breeding ecology of this harrier, and further and more comprehensive studies are required to obtain this vital information.

Having been more than fascinated with the female *Harterti*, that I had the pleasure of observing for two weeks in Mallorca, to see this recognised subspecies in its native environment in Tunisia, North Africa, was an amazing and unforgettable experience. At some stage in the near future, I would like to return to Tunisia or perhaps Algeria or Morocco, the latter two I have yet to visit, to study *Harterti* – in greater detail!

Adult male Montagu's Harrier

Mike Wilkes

4

MONTAGU'S HARRIER

Circus pygargus

Named after the 18th century naturalist Colonel George Montagu (1753-1815), the **Montagu's Harrier** was formerly known as the Ash-coloured Falcon – and is superficially similar to the Hen Harrier. They were so similar, that early ornithologists were unsure that they were a separate species, until the Wiltshire born naturalist (who later moved to Devon), described the smaller species in the *Linnaean Transactions* in 1803.

It is the smallest of Europe's four harrier species and the rarest of the UK's three breeding species, rarely exceeding ten pairs, annually. To everyone's surprise, 2006 was a record breeding year for Britain's rarest bird of prey, with 21 fledged young from 14 breeding attempts nationwide. At the western edge of its range, Montagu's have bred sporadically in the Republic of Ireland during the 1960s (the last record was in County Kerry in 1971). It still occurs in small numbers mainly during the spring migration. In neighbouring Northern Ireland, there were no known records of this species until 29th April 2006, when a 'second summer' male appeared close to the George Best – Belfast City Airport, and remained until at least 3rd May.

No sooner had I mentioned in this chapter, what was purported to be the 'first' Montagu's Harrier for Northern Ireland, which was originally discovered by the late Willie McDowell, when dramatic news of another find came to light and amazingly preceded the 'first' record.

The bird was first observed going to roost in long grass near Ballycarry Bridge close to Larne Lough, County Antrim on the evening of 1st June 2005. Then on the 2nd, Cameron Moore who lives at nearby Whitehead, was informed of a harrier-like bird in the area. He arrived with others, to hopefully secure vital film footage of it. The harrier was a female but at that stage it was not known whether it was a Montagu's or a passing Hen Harrier. Thankfully, and with great difficulty, Cameron managed to film the bird. After a protracted period, which involved many discussions about the bird's origins, the footage was eventually sent to NIBARC (Northern Ireland Birdwatchers' Association Rarities Committee) for assessment.

Thank goodness he did, for in October 2006, and after careful assessment by the committee, it was realised as a female Montagu's Harrier, which now makes it the 'FIRST' for Northern Ireland and not the 'second summer' male which was observed some 11 months earlier in April 2006. Unfortunately the 2005 record was unknown to me personally until quite recently and after hearing about it and receiving a copy of the film from the Records' Secretary, George Gordon, I had no hesitation in including it in this chapter. It is not every day that we see the dainty Monty's here, so two in less than a year are well worthy of inclusion in any book about harriers.

My first ever encounter with a 'Monty's', as they are affectionately known, was during a family holiday to Mallorca in June, 1994. I had previously established a good rapport with Pere Vicens, a warden at Parc Natural de S'Albufera, when we had discussed the resident european Marsh Harrier population and by the second week of the holiday that rapport increased, and Pere confided in me that a pair of Montagu's Harrier were breeding on the island for the first time and amazingly within the Parc! Pere and the rest of the staff were elated because in the past, this species had only been a spring visitor, with only a handful of records in recent years. This pair of Monty's were rightly declared, the premier birds of the year by Pere and his colleagues!

On the evening of 7th June, I made my first attempt at locating the birds. I failed miserably, as I was unsure of the exact area they were nesting. After further consultation with Pere I decided to try again the following morning. At 10.50am on 8th June, I got my first ever views of Montagu's Harrier, as the male arrived with prey and was soon joined in the air by the female. An aerial food-pass then took place close to the nest and after a brief circle directly above it, the female landed complete with prey, which was quite clearly a small bird. The male continued to fly low over the surrounding juncus and tall reeds for five minutes and on one occasion he briefly landed at the nest.

I was treated to excellent views of both birds, with the female in full adult plumage, dark brown and buff above and heavily streaked below. Her prominent white upper tail coverts were clearly visible when in flight. The male however, was not an adult, but a young second-year bird, with virtually no distinct plumage markings. His upperparts were very dark grey, with traces of brown on the upperwings. The upper tail coverts were the only areas that were pale greyish and not defined white. The usually distinctive black bars that are only visible from above and below (in flight) on the secondaries of adult males were absent. The strong sun made it difficult for me to see the male's underparts distinctly, but they appeared much lighter, with a few rufous streaks barely visible on the belly. At the time I summed up his plumage, in just one word – dingy!

Although the Montagu's is the smallest of all the 16 species of harrier, its long and narrow wings gives one an impression of greater size and from a distance (if not experienced on this species' jizz), it would be difficult to distinguish from a Hen Harrier. It has the same general habits of closely quartering the ground, swerving from side to side, and then pouncing suddenly, in a 180 degree turn, on its prey. As the male skimmed over the tall reeds his flight was deliberately slow and uneven, but more buoyant than that of the Hen and resident Marsh Harriers. With the longest wings in proportion to bodyweight of all harrier species, these dainty birds appear very graceful and butterfly-like in flight, as they float over the ground, in an effortless fashion.

The male appeared again at 11.45am, flying low over the reeds in the direction of the nest, and within seconds the female was in the air and had accepted a small prey item, which appeared to be a small mammal, probably a mouse. She hastily returned to the nest with the food and, similar to the first food-pass I had observed earlier, made no attempt to eat it, away from the nest site. As Montagu's are known to be late breeders in the UK and Europe, this female was probably in the early to mid stages of incubation, hence her swift return to the nest. Surprisingly, the nest was situated in a dry area, close to a small bushy hedge and surrounded by juncus and medium-height reeds. The aforementioned were probably 1m high in this particular area and taller elsewhere, with the hedge offering the female a degree of shelter from the scorching sun. Due to the vulnerability of this nest and the first ever breeding attempt for Mallorca, I did not attempt to visit it. It was a great thrill to watch these birds from a discreet distance, rather than causing them unnecessary disturbance and possibly desertion.

I returned again the following day (9th), around noon and stayed hidden amongst the reeds until 3.00pm. Surprisingly, there were no sightings until 2.00pm, when the male, flew directly over the nest, but he was not carrying prey and within a minute he had flown away again. It was a slightly duller day, and his plumage appeared even darker, with a mixture of dark brown and grey conspicuous on the upperwings and tail. To my astonishment the female was not seen and no prey was delivered to the nest during my three-hour visit. This behaviour is not unusual in male harriers. They rarely deliver prey to their partners during the heat of the day, but normal service usually resumes later in the afternoon.

Friday 10th June, was the last full day of our holiday and so I watched the site from 11.45am to 3.15pm. Shortly after my arrival, the male approached the nest from my right, yet once again he was not carrying food for his partner. The curious angle of his approach, made him look like a very large butterfly, as he more floated than drifted, over the reedbed, in a seemingly relaxed manner. He briefly circled over the nest site and then quickly disappeared in a southerly direction. Faint calls were heard, but I was not sure whether it came from him or the female. At 12.35pm the female suddenly took to the air and began soaring in wide sweeps and circles high above the nest site. A small feather fell from her body, and I noticed she had moulted several secondary feathers from each wing.

I briefly observed the male high in the sky directly behind her, but he had disappeared, by 12.40pm. At 12.42pm, the female prematurely cut short her bout of soaring, which was nothing more than a daily exercise flight, and immediately returned to her nest. Minutes later at 12.45pm, the male appeared once again and circled for approximately 20 seconds above the nest, seemingly checking that his mate was OK, but no prey item was noticed in his talons. I observed the male on a further two occasions that afternoon, but not the female, and no food was brought to the nest during my three and a half hour watch.

The lack of regular provisioning was probably due to this young male's immaturity, for over the three days and nine hours of observations, he only delivered prey to the female, twice. A couple of weeks after returning home to Northern Ireland, Pere Vicens, informed me that this nest had sadly failed, the reason unknown. I firmly believe that this nest had failed due to the lack of food being brought to the incubating female, by an immature male, who was probably attempting to nest for the first time. The lack of regular prey deliveries, forced the female to leave her nest unprotected, to try and procure sufficient food, from a male, who was totally inexperienced. I have noticed this behaviour at several Hen Harrier ground nests in Northern Ireland, which also failed, as it is during the incubation period, when most failures occur.

After these unique observations, for me personally, I have certainly a lot to learn about Montagu's Harriers, as they were totally different in their overall behaviour, to that of

Hen Harriers, and certainly not what I had imagined them to be like. In the past, I had to refer to relevant books containing photographs, to gain vital information about this rare species, but nothing beats seeing the real thing! Being the only person other than the Parc staff, to observe and know the location of these birds, was a true testament of their trust in me – Muchas Gracias to everyone at the S'Albufera!

My next encounter with this species, though brief, was during a four-day visit (11th-15th May 1995), to Norfolk in eastern England, with harrier colleague, Roger Clarke. My arrival on 11th May coincided with a Hawk and Owl Trust event which was attended by Her Royal Highness, the Princess Royal. At that particular time, I was the Trust's representative in Northern Ireland and so it was a great thrill for me to personally meet a member of the Royal Family. I also got the chance to discuss my latest Hen Harrier work.

The area Roger and I visited in Norfolk is well known for breeding Montagu's and when we arrived around 10.00am there were at least 20 other pairs of eyes awaiting a sighting of this rare species. The periphery of their breeding grounds was well policed by RSPB warden, Bob Image, who has been responsible for their overall security over many years. We spent around three hours that morning (13th), waiting patiently for a sighting, which I hoped would be an adult male, but sadly none were forthcoming.

Male Montagu's Harrier, Belfast   Derek Charles

At around 4.30pm, we arrived back in the same area, in the hope that we would see at least one of these birds hunting later in the day. To their credit, at least six keen birders were still present, and so eight pair of eyes were obviously better than six, at that time of the day. Shortly after 5.30pm I eventually caught sight of a harrier-like bird in the distance. As it gracefully floated in butterfly fashion over the long grass, I immediately identified it as an adult female. Within a minute, it had discreetly disappeared to presumably hunt elsewhere, as it was not encountered again. When we left the area at 6.15pm there had been no more sightings and I still awaited my first sighting of an adult male Montagu's Harrier!

When Roger asked me to accompany him to India in early December 1997, I considered this, at the time, to be the trip of a lifetime. The Indian Subcontinent is a magical and mystical destination, so to get the chance to visit this vast country for the first time, was an offer too good to refuse. We would be working on behalf of the Hawk and Owl Trust, which ran a joint harrier project with the Bombay Natural History Society. I was assured by Roger, that I would see literally hundreds of male Montagu's

Harriers and at least three other harrier species, over the following two weeks, as we would be visiting at least three destinations which held large numbers of wintering and roosting harriers.

Our trip commenced on 2nd December 1997, arriving in Mumbai (formerly Bombay), in the early hours of the 3rd, the weather was a very humid 27 degrees, which was in total contrast to conditions in England, where it had been snowing prior to leaving London. Around noon we visited Hornbill House, the BNHS headquarters, to discuss our visit with the director, Dr Asad Rhamani and Rishad Naoroji. The remainder of the afternoon was spent sightseeing around this vast city. I could not resist watching hundreds of Black Kites as they swerved and dived through the busy streets, hoping to scavenge anything that was edible.

The next morning we were up early, as we had to catch a flight to Ahmedabad in the Gujarat district of India, followed by a four-hour drive over bumpy and busy roads to our destination at Velavadar Blackbuck Park. The Park is situated on the semi-arid coastal plain, known as the Bhal, to the south of the Gulf of Cambay and is the only open grassland in the State of Gujarat to be conserved as a National Park.

As we were about to enter the Park at around 3.15pm, a male Montagu's in full adult plumage, was observed hunting over an adjacent cotton field. I had waited patiently, for over two and a half years for this moment and nothing was going to spoil it, as I savoured every second of its presence.

In the Park we checked into a room in the Forest Lodge and after a brief rest and a most welcome cup of strong tea, we met up with the BNHS warden N.Ramesh and his assistant Suraj. At around 4.30pm we were driven the short distance by jeep to the main part of the grassland where the majority of harriers roost each evening. The 36sq km area of savannah grassland and thorn scrub was primarily set up to protect the endemic Indian Blackbuck, which was almost hunted to extinction. The Park now holds around 1,000 individuals.

At 5.00pm we took up our position on a high mound close to the road and within a matter of minutes harriers began appearing from all directions. Firstly, several Western Marsh, then large numbers of Montagu's, followed by my first male and female Pallid Harriers. Although the majority were Montagu's, Roger stated that there were an above average number of Pallid present, but very few Western Marsh.

Close to sunset, the constant milling that was observed was an unbelievable sight over the grassland, with at least 1,500-2,000 harriers in the air at any one time – what a sight! By 6.25pm all the birds had gone to ground for the night with the exception of one or two latecomers, who hastily roosted within seconds of their arrival. An extremely late male Pallid flew low over a bush close to where we were standing, inadvertently, flushing six roosting Short-eared Owls. As it had been a long and tiring day, I thought I would probably be counting harriers that night, instead of the usual sheep!

The next morning (5th), we were up bright and early at 5.30am and at the roost by 6.00am to await their departure to the nearby cotton fields, wetland areas and other available foraging habitat. By 6.30am, several of the harriers became very restless, i.e. rising and then roosting again. At 6.40am quite a few birds began rising from the tall grass, briefly circling before flying off in different directions, presumably to hunt. Our vantage point on this occasion was close to where several Western Marsh had been roosting, with two adult males appearing more prominent than the darker plumaged females. By 7.15am, all the harriers had dispersed and so we returned to the Lodge to await the arrival of Bill Clark (US Fish and Wildlife Service), and Vibhu Prakash (BNHS), later in the day. In the meantime we dissected a large number of pellets that had been collected by Ramesh from the roost site, which mainly contained the remains of Tree Locusts and Black spotted Grasshopper, with several whole specimens also available for us to ponder over.

The existence of a large winter roost of harriers in the Park had been known about, since at least the 1980s, and the numbers of birds have peaked at about 2,000 birds (Clarke 1996). In 1997, the numbers of birds roosting at Velavadar were exceptionally high as early as September, and the numbers we had observed during the first few days of our visit were above average for December.

On 6th December, from 4.30pm to 6.30pm, three teams of two individuals each, were positioned on three (south, west and north), of the four sides of the grassland roosting area, where they proceeded to count the birds as they entered the Park. The great majority flew in directly and quite low, with one prolonged and notable concentration of thermalling harriers drifting into the grassland from the southwest, numbering well over 300 birds. The total number of harriers counted, exceeded 2,500. Given that we did not have the manpower to cover the fourth side, and the fact that some birds had already entered the grassland, before we began counting, it seems reasonable to conclude, that more than 3,000 birds attended this unique roost site. The majority were of course Montagu's, but 15-25% was Pallid and a small number, Western Marsh. There were no records of Hen Harriers on this occasion. This roost was three times the size of the largest harrier roosts documented elsewhere in recent times. For example, in Africa (1,000 birds, by Cormier and Baillon, in 1991), and in the USA (1,053 birds, McCurdy et al, in 1995). This site at Velavadar, now appears to be the largest roost of harriers ever documented, anywhere in the world. My memory of that magical and unforgettable evening, when the sky above us was blanked out with harriers is still vivid 10 years later.

After the harriers had departed the roost on 7th December, we walked into a well-used section of the grassland in search of pellets. After an hour Roger and I had acquired a considerable amount, but our search also lead us to discover the half-eaten remains of at least two Montagu's and one Pallid Harrier. Of these, two were reasonably fresh and had probably been preyed upon the previous evening by local Jackals and Indian Wolves, as the harriers roosted amongst the tall grass. The Indian Fox and Common Mongoose were also numerous in this area.

On the afternoon of 8th we returned to Ahmedabad, as the next day we were leaving on an early morning flight to Hyderabad, in southeast India. On arrival we all set off for

Top: Adult male Montagu's Harrier   Bottom: Adult female Montagu's Harrier   *Roger Clarke*

our hotel in Kurnool, which was almost five hours away by car. Then we were on the road again, to Rollapadu Wildlife Sanctuary in the State of Andhra Pradesh, to hopefully observe more roosting harriers that evening. We arrived just in time to see Montagu's and a lesser number of Pallid, milling over two areas of grassland. The light was fading rapidly and so it was impossible to be accurate with our counts, but numbers were only in the hundreds at the most and not thousands as at Velavadar.

The following morning, the 10th, we arrived at Rollapadu by 10.00am and proceeded to walk to the more hilly part of the grassland, where the majority of harriers had been observed roosting the previous evening. To our surprise we found very few pellets in comparison to the number of birds that frequented this area, but once again we found the preyed upon carcasses of at least nine harriers, of which eight were recognised as Montagu's (male and female). The other being a male Pallid. Once again Jackal and Indian Wolf were probably responsible for their demise.

When we walked to where a minority of the birds tended to roost each evening we found a further five harriers that had obviously fallen victim to either jackals or wolves. Four were again identified as Montagu's, the fifth a young male Pallid Harrier. The unexpected discovery of all these dead harriers sparked off a debate later that evening – were they naturally preyed upon, or did they die of pesticide poisoning? There are known predators within the Park at Rollapadu, but we also observed the presence of pesticides being freely used a short distance away the previous day by local farmers. At Velavadar, there are also lots of predators but only three harriers were found preyed upon and no confirmed use of pesticides. Roger Clarke and I firmly believed that these birds probably died of chemical poisoning and were then eaten at the roost by the known predators. Bill Clark and Vibhu Prakash also agreed with this theory.

That evening was also spent at Rollapadu and so we watched the area from a hilly vantage point from 4.00pm until 6.00pm. We estimated that 250 plus birds roosted at both sites, the vast majority being Monty's; the remainder Pallid. The evening of the 11th was our last recognised count at Rollapadu and produced well over 200 birds, but we also observed two wolves in the same area as the harriers, and they were almost certainly the predators responsible for the predation of the nine harriers that we had found the previous day. During the winter of 1985-86, 800-1,000 birds were recorded here during survey work to assess the population of Great Indian Bustards. Sadly numbers have dramatically declined since then. Rollapadu was another superb site to visit, with three Great Indian Bustards being one of the many highlights.

On 13th December, we would be embarking on the final leg of our journey via New Delhi, which would eventually take us to the most famous wildlife sanctuary in India – Bharatpur! At 3.45pm, on the14th, we made our way to the grassland harrier roost site within the park and when we left at 5.45pm, there were around 40 harriers present comprising three different species, all of which are described in the individual species' accounts elsewhere in this book. There were however, no Montagu's in this gathering.

Late in the evening of 15th, we flew back to London and I arrived back in Belfast the

Arrival of a large roost of Montagu's, Pallid and Western Marsh Harrier in north-west India   *Roger Clarke*

following morning. What a 'harrier' holiday in India it had been and a return visit was scheduled for the near future.

In late November/early December 1999, Roger and I, together with Graeme Hewson and David Miller, re-visited Velavadar in Gujarat, northwest India. Driving to the Park, from the city of Ahmedabad, it was plain to see that 1999 had been a severe drought year with much of the cotton and other crops looking somewhat pitiful and sparse in the nearby fields. On the first two days of December, we counted the outgoing and incoming harriers on various sides of the Park, each morning and evening. On the assumption that similar numbers of birds used the same sides on successive days, we came up with a figure of approximately 1,100 birds. The Forest Department's coordinated count on 28th November prior to our arrival, gave a figure of 1,075 birds – about one-third of our record December 1997 count.

Severe fluctuations in numbers from season to season are to be expected in wintering quarters, where distribution is largely dictated by feeding resources. About 60% of all the harriers were grey males and of the 1,100 birds present, roughly two-thirds were Montagu's and one-third Pallid. In this area Montagu's mainly prey on locusts each winter but as there were too few crops to harbour them, due to the drought, food was obviously minimal, hence the reduction in their overall numbers. After a full week of observations, where numbers remained relatively stable, we moved to pastures new, in the hope that the heady harrier days of 1997 would return to Velavadar in the near future.

By mid morning on the 7th, we had left Velavadar, and after a long and tedious six hour journey (travelling northward), we arrived at Dasada, to meet up with our hosts for the next four days, Sarfraz Malik and his family. The purpose of this visit was to monitor roosting harriers, in an area known as the Little Rann of Kutch. The Little Rann is situated in the Thar Desert of north-western Gujarat State and the Sind Province of Pakistan. This vast area comprises 4,950sq km between the Gulf of Kutch and the mouth of the Indus River in southern Pakistan. The Rann is more famous as a Wild Ass Sanctuary, than for roosting harriers in the desert, so it was expected that this would be an interesting visit.

Roger Clarke had been taken to this area in January 1993, to watch a small communal roost of mainly Montagu's Harriers (and two other harrier species), utilising the bare open ground on most evenings. During one evening, (22nd), he observed around 20 birds and probably over 30 on the next evening. These numbers may have been even greater but he found it difficult to assess owing to the poor light as dusk fell. Now, I too was about to witness this unusual form of harrier roosting.

On the evening of the 8th, we observed at least 12 birds roosting on open ground, with others seemingly travelling on to occupy ground further east. On the 9th there were around 20 harriers, with at least a further 10 roosting elsewhere. Our final watch on the 10th saw similar numbers to the previous evening, but this time they were more scattered and difficult to observe clearly. Shortly after sunset each evening we drove around to try and find their exact location and other small roosts, but without success. Certainly this form

Top: Exquisite dark-morph Montagu's Harrier   Bottom: Female Montagu's Harrier   George Reszeter   W. S. Clark

of roosting not only raises some doubt as to whether harriers choose roost sites solely for shelter, but lends weight to the theory, that these are sites where harriers can hear well, especially when ground predators are active at night. For the record, the majority of the birds observed were mainly Montagu's, with a few Pallid also present.

In early December 2000, we set off on our next harrier trip to India, the third visit in the past four years. On our way to Velavadar from Ahmedabad, we again encountered fields upon fields of crops (mainly cotton), which had simply died, due to the lack of rain and watering by the local farmers. This was the second consecutive drought year we had been faced with and so this probably meant we would not be observing large numbers of harriers at the grassland roost within the Park.

That evening, the 7th there were only around 200 harriers present, mainly Montagu's. Sadly, large numbers failed to materialise during the four days we spent in the Park. This was most disappointing, but not totally unexpected, as when such conditions prevail, there is a shortage of prey. I must say, however, that the number seen was 194 more than what I would observe at my local roost in Northern Ireland, and so I was not too disappointed.

On the 11th, 12th and 13th, we visited for the first time within the State of Gujarat, an area known as the Royal Grasslands, at Umwada, a short distance from the city of Rajkot. These huge areas of grassland were owned by the local Maharajah and the Maharani (for cattle grazing), whom we had met on return from Velavadar. They kindly gave us permission to visit these areas, as harriers had been observed there in recent weeks. They also supplied us with a driver and jeep, which was most generous and greatly appreciated. Over the three evenings, counts of only 10, 12 and approximately 13 harriers were observed – mainly Montagu's. There were however more larks present here than at Velavadar, so why were there not more harriers? This part of Gujarat did not appear to be affected by the drought, but perhaps the majority of all three regularly seen species had moved further afield, where prey was more easily obtainable.

During a family holiday to Kenya and Tanzania, in late March/early April 2002, my main interest was focused on observing as many species of raptor as I could and of course the fantastic array of wild animals that were also freely available. But it was when we moved from Mombasa to northern Tanzania, visiting world famous sites like Tarangire, Ngorongoro Crater and the vast plains of the Serengeti, that I became aware of the large numbers of Montagu's, that were present at the latter location.

On our way from Arusha to Tarangire, only a few individuals were observed patrolling the lush green fields that would have been more akin to Northern Ireland than Tanzania. When we were exploring the vast Ngorongoro Crater (2,286m asl.), we noted much larger numbers of Montagu's, mainly grey males and only an occasional female. On 31st March, when travelling along the dusty tracks of the Serengeti, the most common raptor was surprisingly, Montagu's Harrier! The grass here was extremely long due to a prolonged rainy season and during several 'Big Game' drives, large flocks of larks and finches were flushed from the tracks and the adjacent grassland. This whole area was ideal hunting and roosting habitat for Montagu's Harriers.

That evening, around 6.15pm, and approximately 3km from our lodge, ten Monty's were seen circling low over the grassland, close to a tall Acacia. I immediately begged our driver to stop and by 6.30pm, the numbers had increased to 27. When we had to leave at 6.40pm, there were at least 35 birds present and roosting was now taking place, the majority being grey males. I must stress, that it is illegal to remain on the Serengeti after sunset for obvious reasons, so I sincerely thanked our driver for being so helpful and obliging.

The next evening, as we approached the same area at 6.00pm, up to eight Monty's were pre-roosting on and by the sides of the track. A grey male, which was inadvertently flushed by our jeep, dropped a small prey item and when cautiously retrieved by our driver, it was a partially eaten locust, which was similar to the ones collected at Velavadar, in India. By 6.30pm, over 40 birds were in the air with several pre-roosting on the large Acacia. Sadly, we had to leave at this important stage of the proceedings, but the next morning (2nd), I managed to catch two of these locusts, close to our lodge, which I duly kept as specimens for my own reference.

There were probably other roosts waiting to be discovered on these vast plains, as it was common for us to observe 100 plus birds, during our daily forays to different parts of the Serengeti. It was obvious from the large numbers we encountered daily, that not all had migrated north to their main breeding grounds in the UK and various parts of Europe, but this was probably imminent. As mentioned earlier, the majority of Monty's, we observed during our two-week safari, were grey males, whereas in India, females and ringtails, dominated the roost and daily sightings. Not since my three visits to Velavadar, have I observed so many Montagu's Harriers in one area.

I also observed Montagu's on the Masai Mara game reserve, in late June/early July 2001 at Tsavo East, in Kenya. On both occasions my records involved wandering individual males, which may have been non-breeders, with no urge to leave these prey-rich areas. Having now seen literally hundreds and probably thousands of Monty's, over the past ten years, I have yet to encounter a bird in the dark morph (melanistic) plumage. This rare plumage phase is apparently more common in the west of its range ie in Europe, where birds are regularly observed in France and Spain and also on passage. Further east, although rare, they occur amongst birds that are found wintering in India, western Asia and also in eastern Europe. Hopefully, I shall find one in the near future, which would then fulfil my studies of this delicate looking little harrier.

Male Pallid Harrier

Roger Clarke

5

PALLID HARRIER

Circus macrourus

The **Pallid Harrier** or Pale Harrier is a very rare vagrant to the UK and western Europe. It is a bird of the Asiatic Steppes and open plains of eastern Europe and central Asia, breeding primarily in Russia, Kazakhstan and northwest China. Small populations also breed in Azerbaijan, Romania, Turkey and the Ukraine, with occasional irruptive movements producing breeding records as far west as Sweden and Germany. Winter quarters, for this migratory raptor are mainly in India, Pakistan, Sri Lanka and the continent of Africa, south of the Sahara, but a minority overwinter on the edge of their breeding range, with birds also recorded in Cyprus, Italy, Greece, Israel and Egypt. A first-winter male, has even been recorded for the Seychelles in the western Indian Ocean, on 25th January 2002. Due to serious threats in it's mainly grassland breeding grounds, this harrier was classified by Birdlife International in 2006, as – near threatened!

We in Northern Ireland and Ireland as a whole, live more in hope than expectation, that rare birds like the Pallid Harrier, may occur one day on our small westerly island. But when our first ever Montagu's Harrier suddenly appeared in Northern Ireland, my doubts were suddenly replaced by renewed optimism, for the appearance of the Pallid Harrier here, one day.

Up to and including 1952, there had only been three records of this species in the UK, but since 1993 this has now risen to around 20 individuals. Historically, the May 1993 record in Tayside, Scotland is perhaps significant, in that a male exhibited territorial behaviour and then began displaying and calling to a female Hen Harrier. This was surpassed on the island of Orkney in May 1995, when a male Pallid was observed sky-dancing and copulating with a female Hen Harrier. Consequently, a nest was built and five eggs had been laid by 2nd June, but on 22nd June, the nest was found to be empty, with the eggs probably predated by a Hooded Crow, which are a common cause of nest failure in Orkney. This 1995 record, is the first authenticated record of mixed pairing between Pallid Harrier and Hen Harrier and only the second ever incidence of mixed pairing between Pallid and any other harrier species. The previous record of a successful mixed pairing was in northern Finland in 1993, when a male Pallid Harrier bred with a female Montagu's Harrier and raised three young. For those of us studying the genus – *Circus*, it is well worth remembering that Pallid Harriers are a very early spring migrant, so any odd-looking male Hen or unseasonable Montagu's, could well be – a Pallid!

Until I visited India for the first time in December 1997, I was not familiar, nor had I ever observed this species in Europe. Therefore, my knowledge of Pallid Harrier was somewhat limited until I arrived on the Indian Subcontinent. My first ever Pallid Harrier, a striking male, was observed at Velavadar National Park, in the State of Gujarat, in north-west India, from the balcony of my lodge within the Park, during the afternoon of 4th December 1997. I was enjoying a welcome cup of strong Indian tea at the time, before leaving to observe the multitude of harriers that would grace this Park later that evening.

While in the company of Roger Clarke, I observed this male dashing across the grassland in pursuit of a mixed flock of larks, in an unharrier-like manner. Yes, their hunting flight is typical of other harrier species, but it is much faster and falcon-like when prey is located and about to be seized. On this occasion the bird was unsuccessful, but it was certainly a learning experience for me, observing a Pallid hunting in what I thought was an uncharacteristic manner. What I did clearly notice was, that when all three species arose from the roost at sunrise each morning, Montagu's and Western Marsh, departed immediately to hunt outside the Park, while the majority of Pallid Harriers remained in this vast grassland to hunt the large flocks of small birds that were abundant here in 1997. During the course of my visit around 20% of the roosting harriers at this site were Pallid, with males, females and juveniles, regularly seen throughout the day and in the evening prior to sunset.

When it came to identifying female Pallid from female Montagu's, I had real problems! Both look similar in the field, particularly in flight, but it was only when they pre-roosted on mounds and open ground, could they be separated, but with great difficulty in my case. The legs of Pallid are much longer than that of the Montagu's Harrier, but their wings are much shorter and fall short of the tail tip when perched, whereas as the wingtips of the latter species are much longer and reach the tail tip. Thankfully, there were no female Hen Harriers present, as they too can cause identification problems to novices like me. Juveniles of both Pallid and Montagu's are also similar and show unstreaked rufous underparts, but adult male Pallid Harriers are among the palest of all raptors and, from a distance they can appear all white except for a small black wedge in both wing-tips. The upperparts are very pale silvery grey and again show the distinct black wing-tips, but the uppertail coverts do not show a distinct white patch as on most species of harrier. It was an absolute thrill for me to absorb and observe this species for the first time and thankfully they were not my last sightings before I went home.

A few days later, I was savouring more roosting Pallid Harriers at Rollapadu Wildlife Sanctuary in Andhra Pradesh State, southeast India. Pallid numbers here were greatly reduced from that of Velavadar, and of the 200 plus birds present from 9th-11th December, only five per cent of them was of this species, the greater number being Montagu's. Then before departing for London, Roger and I spent our last two days in India at Keoladeo NP, north of the capital New Delhi, which is better known to most British birders as – Bharatpur. Only 14th December was spent watching the small harrier roost there and that evening, out of 40 or so birds of three different species that utilised the site, four were Pallid, of which only one was an adult male, the other three being adult females.

When I returned to India in December 1999, very small numbers of this species were observed at Velavadar and elsewhere, due to the drought conditions that prevailed. Numbers at the roost had decreased to 1,100 birds, of which only one-third were Pallid, the remaining two-thirds were typically Montagu's. Further north in Gujarat State, on 8th, 9th and 10th December, while visiting the unique desert roost in the Little Rann of Kutch, only a handful of Pallid Harriers were observed over the three evening watches,which was most disappointing.

Juvenile Pallid Harrier at Tronda, Shetland in 2004

Hugh Harrop

Similar conditions were again witnessed at Velavadar in 2000; only this time numbers had hit rock-bottom, with probably no more than 200 harriers present. Of these, only around 50 were definitely Pallid Harriers and in this huge grassland, where in the past these birds were regularly observed, preying upon the large mixed flocks of larks and finches, very few small birds were present. Obviously the number of harriers present at Velavadar is inextricably linked to the amount of prey available – no food, no harriers! Within the same State, things were not much better at the Royal Grasslands at Umwada, with only a few Pallid amongst the majority of Montagu's.

I had to wait until 2002 to observe this species again, but this time it was on the grassy plains of the Serengeti in Tanzania. Only a few males were observed here, but I also got the chance to watch two of them hunting small birds, which were particularly numerous in the area. They hunted in true harrier fashion, quartering slowly in tern-like flight over the long grass until prey was detected, then they would wheel in a sudden 180 degree turn, with both feet thrust downward and strike out at their quarry with great speed and agility. Both males I had been observing were successful, with one catching what I determined was a lark, with the other flying off with what appeared to be a small lizard.

My next encounter with a Pallid Harrier, occurred on 16th July 2004 when on a family holiday to Tsavo East, in Kenya. While taking a welcome break close to the River Galana, five harriers suddenly appeared from virtually nowhere and began hunting close to the river's edge. When I finally trained my binoculars on them, all five were males, with three Montagu's and two Pallid. I never really discovered what they were hunting along the sandy bank of the river, but it may have been lizards, which were probably lazing in the hot mid-day sun. A short time later I encountered both Pallid Harriers, which were actually second-winter birds, as both showed brownish-grey markings on the head, back and upperwings, as they lazily weaved their way over the grassland. At this particular time of the year adult males and females would be on their breeding grounds in northern and eastern Europe, so these two were obviously non-breeding birds.

It must be said that Pallid Harriers and similarly Spotted Harriers have strong nomadic tendencies to seek out new breeding areas, which are usually distant from those previously occupied. If in spring the latter are unsuitable, mainly due to fluctuations in Lemmings, voles and other small rodents, then Pallid Harriers will seek out new and suitable areas for adequate food. Due to this irregular behaviour, particularly during most breeding seasons, it is difficult to estimate the total population and nesting habits of this species, as there is no guarantee that males and females will return to the same area each year. Therefore, from my point of view, it is probably best to study this eastern wanderer, on its breeding grounds in the vast Russian Steppes, rather than in wintering areas where little can be learned of its true lifestyle and ecology.

With more and more individuals seemingly visiting the UK on an almost annual basis, who knows what the future holds for the Pallid Harrier, especially in western Europe. It is not surprising that the majority of birds found in western Europe tend to be adult males, given that females and juveniles are much harder to identify!

Top: Juvenile Pallid Harrier   Bottom: Female Pallid Harrier with a Lesser Bandicoot Rat   *Roger Clarke   Jugal Tiwari*

A late but important addition to this chapter highlights the sheer brutality shown to this species on the island of Malta on 2nd April 2007. According to Birdlife Malta, a male Pallid Harrier was shot on migration in the south of the island, a week before the start of the controversial spring hunting season. It had gunshot wounds to both wings and chest; with one wing so badly injured that bones could be seen sticking out through the feathers.

In Europe the Pallid Harrier is now becoming an extremely rare species and although protected under EU Law, numbers are still declining throughout much of its breeding grounds. The European population is now estimated at only 300-1,200 breeding pairs (Birdlife Malta), and if one excludes its main breeding grounds in Russia, then the rest of Europe has a meagre 50 plus pairs, maximum. In light of this most recent atrocity and other well-documented carnage to birdlife over several decades, particularly during the spring and autumn migration, Malta is not a country that I shall be visiting in the near future, or ever if possible!

Adult female Pallid Harrier

Black Harrier in flight   *Andrew Jenkins*

6

BLACK HARRIER AND AFRICAN MARSH HARRIER

Circus maurus and Circus ranivorus

The **Black Harrier** has the most restricted range of any continental species and is endemic to the grasslands, fynbos and karoo, of southern Africa. The world population is estimated at fewer than 1,000 birds, of which, less than 100 occur in protected areas, and as such is now classified as locally – near threatened and globally vulnerable!

The **African Marsh Harrier** as its name suggests, is a bird of wetland areas, but in parts of southern Africa, its numbers have been declining due to the drainage of marshes and swamps for agriculture and development. The population in southern Africa is estimated to be only 3,000-5,000 pairs (possibly lower) and, like the Black Harrier, it has recently been elevated to a 'species of conservation concern'!

The huge continent of Africa and its related islands hosts approximately 110 species of diurnal raptors and 45 species of owls, many of which breed. Others use the vast grassy plains, wetlands and mountains as a winter staging-post due to the diversity of prey available to them. Taking into consideration that the Indian Subcontinent can host up to six species of harrier each winter, this number can be surpassed by a further two in Africa if one includes the subspecies *Harterti* and an island species in the form of the Madagascar Marsh Harrier. A breakdown of all eight species shows that Hen, Western Marsh, Montagu's and Pallid only winter there, with the remaining four, African Marsh, Black, Madagascar Marsh and the recognised subspecies *Harterti,* being regular breeders. Of those mentioned, my particular interest would be in studying the African Marsh and the endemic Black Harrier.

The year 2000 saw the start of a research project on the biology and conservation status of the Black Harrier in the Western Cape, initiated by staff and associates of the Percy Fitzpatrick Institute of African Ornithology at the University of Cape Town. Incidentally, the Black Harrier has been the Institute's logo since the early 1980s and this new initiative was the first formal attempt to understand this harrier's breeding biology.

The co-ordinator and principal fundraiser, Dr Andrew Jenkins, firmly believed that given sufficient funding and enthusiasm, this worthwhile project could be expected to last at least five years, so that a true picture of this rare and endemic harrier's lifestyle could be finally understood. Leading the research team for Andrew Jenkins, was my harrier colleague and author, Dr Rob Simmons, Odette Curtis and Chris Rhodes, a research student (from Edinburgh). Odette is probably the only female Harrier researcher in South Africa and on the continent of Africa as a whole!

Similar to Northern Ireland, there is a dearth of raptor workers in the Western Cape, so when my Hen Harrier season ended in late July 2002, I had no hesitation in volunteering to help my fellow harrier colleagues in South Africa during my 12-day visit there.

On 21st September 2002, I made my way to Dublin Airport and from there departed for London Heathrow, followed by a 12-hour overnight flight to Cape Town. The purpose of this visit, which I had first contemplated making in 2000, was to participate in the ongoing breeding survey of the endemic Black Harrier and also to observe the smallest

Female African Marsh Harrier with chicks

Peter Steyn

of all six recognised marsh harrier species, – the African Marsh Harrier! Incidentally, there are no plumage differences between sexes of the Black and African Marsh Harrier and similarly the Spotted Harrier from Australia, with females of all three species somewhat larger. Of the 16 recognised species worldwide, these are the only three, with identical plumages!

Prior to my arrival in Cape Town, I had already agreed with Andrew to monitor breeding Black Harriers, in the West Coast National Park (130km southwest of Cape Town), which to my delight was a stronghold for both species. I was based in the picturesque holiday resort of Langebaan which was convenient to one of the Park's main entrances.

On arrival in Cape Town on the 22nd, it was necessary for me to hire a small car at the airport for the duration of my visit, but finding my way to Langebaan was a tricky ordeal to begin with, due to the heavy traffic and not immediately finding the correct road. Then, after several wrong turns and an unscheduled tour of Cape Town, I eventually arrived at my destination around lunchtime, where I duly relaxed in my hotel for the remainder of the day.

The following morning, I eagerly entered the Park at 9.00am, and proceeded to drive through unfamiliar harrier habitat. I was now viewing for the first time fynbos, which is a mixture of small bushes, scrubland and grass, with an array of wild flowers adding a certain beauty to the whole area. To my astonishment, the Park (27,000ha), was almost devoid of trees, with the adjacent Langebaan Lagoon, providing a seaside backdrop to the whole area.

At exactly 9.18am, came my first Black Harrier sighting, as one flew extremely low across the road in typical harrier fashion, in front of my car. 'What a beautiful bird', I quietly remarked, as it drifted away to my left! Not surprisingly (though I am obviously biased), the Black Harrier is considered by many to be South Africa's most attractive bird of prey and rightly so! This sighting was quickly followed at 9.30am, by a second individual, with a third bird more closely observed at 9.45am. This particular bird, gave me superb views of its pied plumage and foraging technique amongst the density of the fynbos, but surprisingly no prey was obtained after several unsuccessful strikes at probably a small mammal or bird. These birds are obviously highly skilled at locating and capturing prey amongst the dense undergrowth that exists here, but when their intended quarry moves deeper into the vegetation, it was immediately noticed that they struggled to penetrate any living form and quickly moved on to another location. The above was probably an isolated incident, for during my stay in the Park, I observed at least two kills per day, with all the prey items successfully retrieved from the dense swathes of fynbos etc, proving that these harriers were exceptional predators in a unique habitat. During the course of the day, I regularly observed at least 10-12 Black Harriers (not including recently fledged juveniles), and then I observed my first African Marsh Harriers, which later appeared just as numerous as the aforementioned species.

It was at Geelbek I encountered these birds, as they regularly patrolled the huge

Black Harrier at the nest

Andrew Jenkins

saltmarsh and an extensive reedbed in search of prey. Geelbek is regarded as a mini stronghold for African Marsh Harriers, so to have both species nesting side by side was an added bonus for me personally. Several spectacular food-passes were observed immediately above their respective nest sites, with males and females delivering food to nests which clearly contained hungry chicks. Occasionally, birds were seen hunting over the fynbos, close to my roadside observation point, but the majority of their foraging was concentrated over the reedbed and along the periphery of the saltmarsh.

During my daily visits to this their main breeding area, it was not uncommon to observe up to 10 birds in the air simultaneously, as I recorded vast amounts of prey and even nesting material being brought to the nest sites. A favourite area for several of the food-passes was along a grassy bank adjacent to the reedbed and when one or other of the adults consumed the prey item, the birds would fly over the fynbos, within metres of my car (which I used as a hide), before returning to their nests. Consequently, I obtained superb views of their mainly brown and rufous plumage with heavily barred underwings, as they lazily drifted past my vantage point. The size of these individuals was also noted and being obviously smaller than their five much larger cousins, they appeared more graceful and buoyant in flight and seemingly more agile.

South African raptor doyen, Peter Steyn, told me recently that the African Marsh Harrier is very dear to his heart, as it was the first raptor he studied after the Red-breasted Sparrowhawk. This was during the 1950s when one could easily find several nests in a weekend at Rondevlei Bird Sanctuary, close to his home at Newlands. As a result of urban sprawl and habitat loss in the Cape Peninsula area, it is now considered a rarity there. Thankfully I received better news from Peter towards the end of July 2007, when he observed a pair holding territory at the Sanctuary, so here's hoping for a successful outcome! Rob Simmons has also reiterated Peter's sentiments about this species, stating that they are rarely seen these days and we do not really know why they seem to be in a steep decline. Perhaps it is not just habitat loss that is responsible for its demise.

On the 24th, I was joined in the field by Andrew, who had kindly driven up from Cape Town to meet me at my new accommodation around 8.30am. Then, armed with a permit and key, we spent the morning exploring and getting me familiarised with harrier life in the West Coast National Park. Although no specific nest sites were visited (this would occur on 26th), we did frequent the periphery of the Black Harriers' breeding grounds, in an area that was not accessible to the public. I was amazed to hear that 11 pairs were nesting in this, the eastern section of the reedbed, and almost adjacent to where the African Marsh Harriers were nesting. It was true to say that the Black Harriers, were infringing on Marsh Harrier territory, but to my surprise no interaction was noted between either species, especially when two African Marsh Harriers, repeatedly frequented the former's breeding grounds during our stay. I expected the Black Harriers' nests, to be in dry areas and not amongst juncus and wet marshy ground, but Andrew assured me this was common here, with the aforementioned habitat also regularly utilised.

During our visit, we observed up to eight Black Harriers, including a recently fledged

Black Harrier chick and unhatched egg   *Rob Simmons*

juvenile, which sported dark brown patches on its upperwings and definitive black and grey barring on the tail, although not as clearly defined as that of an adult bird. At 12.45pm another fledgling flew from a small bush close to the path to receive food (which I was reliably informed, was a Striped Mouse), from a parent bird. The Striped Mouse accounts for 68% of the Black Harrier's diet, with the remainder being, small birds (30%) and reptiles (2%). The resident African Marsh Harriers also rely heavily on Striped Mice and Vlei Rats, with the former accounting for around 70% of their total diet. Before leaving to go elsewhere, I remarked to Andrew that I found it hard to determine in these two harrier species, which was male or female, as their plumages are identical. Obviously size is a tell-tale factor when both sexes are observed close together in flight, the female being the larger of the two.

We left the park travelling south to other known Black Harrier territories near Yzerfontein. Unfortunately both nests (amongst juncus, in dry ground), were unexpectedly deserted and so we moved on to Koeberg Nature Reserve, which was approximately 35km from Cape Town. At this location we observed four adults and two recently fledged juveniles, one of which was obligingly fed with a small mammal, close to our observation point. A female with an orange wing-tag and a male with a white tag (on their right wings only), were close by, with the former vigorously sky-dancing, which was more a territorial display, due to the intrusion of this male.

Andrew then showed me two recently vacated nests which still had fresh excreta and down, clinging to the surrounding vegetation. These two nests were in short grass (known as Red Grass), amongst stabilised sand-dunes, which were in total contrast to the habitat utilised in the WCNP. To complicate matters even further, Black Harriers also nest in montane regions (similar to Hen Harriers) at Slent, to the east of Koeberg, which was again in contrast to the habitat I had been shown earlier that day. As it was now late afternoon, I thanked Andrew for his valuable help and guidance during our fascinating look at Black Harriers and their contrasting habitat and lifestyle and slowly drove back to the Falcon's Rest in Langebaan.

I spent the following day (25th), observing both harrier species in the WCNP and when driving near Geelbek, I discreetly watched an African Marsh Harrier, feeding furiously on a recent road-kill, which I later identified as a Cape Francolin. A short time later, the same bird was seen hunting close to my car and as it flew off towards the nearby marsh, I noticed it was carrying a frog in its talons. Incidentally, the Latin name for this species is, *Circus ranivorus*, which means, frog-eating harrier, which is misleading as only 2%, of its diet consists of frogs! During my morning observations, at least three African Marsh Harriers, regularly foraged along the periphery of the reedbed, with prey in the form of Vlei Rats and Striped Mice, brought back to their respective nest sites. As mentioned earlier, carrion is also an important part of their diet and in the past I have seen amazing photographs of African Marsh Harriers, standing in water consuming fish, which is more akin to the resident African Fish Eagles and Ospreys, than harriers. Before leaving Geelbek, I watched intensely a Marsh Harrier seriously mobbing a Marsh Owl,

Young African Marsh Harrier chicks

Peter Steyn

which suddenly arose from the reedbed. The latter fled to the safety of the tall Phragmytes Reeds, only to be attacked by a Black-shouldered Kite.

The 26th heralded the arrival of Rob Simmons and Chris Rhodes, who unexpectedly arrived at my accommodation before 8.00am. The majority of the day was spent visiting Black Harrier nests in the wet areas dominated by juncus and sedges, with nests occasionally located close to or under small bushes. The 11 nests occurred in an area of 0.67sq km with an average 'nearest neighbour' distance of 340m. All appeared to be monogamous and no aggression was recorded between the breeding pairs. Rob commented that this close nesting confirmed their semi-colonial breeding habits.

Amazingly, all 11 nests were successful, with over 30 chicks fledging from all the known sites. During the day we visited six nest sites, which had previously been monitored during the egg stage. At these sites I helped Rob and Chris with the duties associated with recording biometrics and ringing of chicks, with one nest containing an unexpected brood of four. According to Rob, two or three chicks is generally the norm for Black Harriers, four is exceptional. When visiting another nest, the female appeared reluctant to leave her three chicks, and only when we were within a metre of her did she decide to leave. Rob explained to me later, that given the extraordinary tameness of some incubating and brooding females, it is possible (occasionally), to walk discreetly up to the nest and catch them, if need be. To achieve this successfully, you require a keep net, complete with a long pole, with apparently one female caught in every four attempts. This is probably the only species of harrier that will allow you to get this close, when visiting their nest sites.

During my daily forays into the park, I managed to find another two pairs of Black Harriers which were observed sky-dancing and also copulating. These, I informed Rob and Chris about before they left for Cape Town, but as it was late in the day, they decided to make a return visit later in the week. I had first observed these birds on 23rd September and they were still present on 25th, prior to Rob and Chris's arrival on the 26th. They apparently had no record of Black Harriers in the area that I had roughly described to them, so these were probably two new and additional pairs.

On the 27th, 28th and 29th September, my attention was focused on an area of the Park that was a considerable distance from the main core area. Each day I visited the area to monitor these birds, and then in the evening I would inform Rob of their progress and more importantly confirm that they were still present.

On the 30th, Rob and Chris made a welcome return to the WCNP, and so I took them to the western side of the lagoon, where I had diligently watched these birds in their absence. This new drier site on the road to Patternoster was amongst scrub and sand-dunes, but after extensive searches no nests were found, despite the four birds still holding territory. About mid-afternoon, Rob and Chris bade me farewell, with the promise that they would return to this area again in the hope of confirming another two breeding pairs. I continued monitoring this particular area until I left for home on the morning of 2nd October. The two pairs were still vigorously displaying, followed by the occasional food-pass and then short bouts of copulation. Even nesting material was carried in by one prospecting male. I believed these birds were possibly late nesters, but unfortunately I would not be around to observe the outcome. Just to add, that when I returned home, Rob informed me that he had visited the area on a further two occasions, without success.

During my visit to this unique habitat, I had observed many aspects of Black Harrier ecology, which included sky-dancing, food-passes, the carrying of nesting material, repeated copulation, nests with eggs, nests with recently hatched chicks, nests with fully feathered juveniles, and recently fledged birds. Similar observations were recorded during my regular sightings of African Marsh Harriers, near Geelbek, although sadly I did not record a dark morph for this species, which does occur occasionally.

This was a most memorable trip, which I would not have missed and I hoped that it would be possible for me to visit the West Coast National Park again, sooner rather than later, in my lifetime. To get the chance to handle and work with breeding Black Harriers, under the guidance of Rob Simmons, was a phenomenal experience to take back to Northern Ireland. This was the sixth species of harrier, I studied and it continues to remain one of my personal favourites. What a beautiful and sensational bird to behold in the field!

African Marsh Harrier at the nest

Peter Steyn

Male Pied Harrier

7   PIED HARRIER AND EASTERN MARSH HARRIER

Circus melanoleucos and Circus spilonotus

The **Pied Harrier** breeding strongholds are in eastern Russia, north-eastern China, Mongolia and northern Burma (now called Myanmar). The species has also recently been confirmed breeding in the foothills of the Bhutan Himalayas and Assam, in northeast India, but only in small numbers at the latter two locations. Northern and eastern India, Pakistan, Myanmar, Thailand and also the Philippines are regarded as wintering strongholds for this attractive species.

Rumours abound that this species has bred in the Philippines, but there has been no formal evidence to confirm this.

The **Eastern Marsh Harrier** breeds in the grasslands and wetlands of southern Siberia, northern Mongolia, northeast China, Manchuria, Japan and probably North Korea. In eastern Asia, the Western Marsh Harrier is replaced by the Eastern Marsh Harrier, but occasionally both species overlap and hybridize on the eastern shores of Lake Baikal. Wintering birds have been recorded in India, Pakistan, Myanmar, Thailand and the Philippines, with birds migrating as far south as northern Borneo.

There have been several unsubstantiated records of this species visiting mainland Australia, which to date remain problematical and have not been accepted. Confusion also exists, as Eastern Marsh Harrier males and females have similar plumage detail to that of the neighbouring Papuan Harrier, which is apparently sedentary on New Guinea.

The greatest diversity of breeding harriers in the world is found in eastern Europe, in an area near Kiev, western Russia, where one could theoretically find Pallid, Montagu's, Western Marsh and Hen breeding together. However this diversity is surpassed on the Indian Subcontinent, where wintering harriers can include six species with Pied and Eastern Marsh being added to the four already mentioned. During my three previous visits to India (1997/99 and 2000), and during casual conversations with several other raptor enthusiasts (local and visiting), it was mentioned that it was possible to encounter all six species in Kaziranga National Park, in far-off Assam.

The state of Assam lies astride the mighty and revered Brahmaputra River, in the remote northeast corner of India. The world famous Kaziranga National Park in the Nowghan and Golaghat districts of central Assam, covers an area of 430sq km, and is the largest protected area on the southern bank of the Brahmaputra River. The park extends from the river in the north to the National Highway (No 37), which runs along its southern boundary at the foot of the Karbi Anglong Hills. The terrain of the park is flat (55-75m asl), with an east to west decline and is therefore ideal for wintering harriers.

The Park is the ancestral home of the protected One-Horned Rhinoceros, with large numbers of Asiatic Wild Buffalo, Tiger and Indian Elephant, also roaming freely. The vast landscape of Kaziranga is therefore a wide variety of habitat with sheer forests, tall Elephant Grass, rugged reedbeds, mellow marshes and shallow pools, scattered over a large and unspoilt area. Around the periphery of the Park, are huge rice fields, lush green grassland areas and bordering tea plantations, which would hopefully give me superb views of Pied and Eastern Marsh Harriers and any other *Circus* species.

Female Eastern Marsh Harrier

Romy Ocon

So, going on the information I had already received about this area, I set off on 13th November 2003 for a week long visit to Kaziranga National Park to study wintering Pied and Eastern Marsh Harriers. From the start, it was not an easy destination to get to, via London, Delhi and then Guwahati, followed by a six-hour (248km) drive to the idyllic Wild Grass Resort, which is situated close to the Park. On arrival, in time for dinner on the 14th, I was totally exhausted and retired early, as I would be making a prompt start the following morning.

At 7.00am on the 15th, I was driven by jeep, complete with guide, to visit the Central Range. The Park opens at 7.30am until noon and then from 2.00pm until 4.30pm. On arrival, it is obligatory to acquire an armed guard, because of the danger of attack from any of the aforementioned wild animals as well as Leopard and Jungle Cat, which probably offer a lesser threat. Within the Park you are not allowed to stray far away from your vehicle, for obvious reasons. Three of these large animals, Rhinocerous, Buffalo and Elephant were much to the fore during this visit and I soon realised that this was not the habitat for foraging harriers, mainly due to the extremely tall grasses and virtually no wetland areas.

The afternoon visit was spent in the Western Range and it was here that I saw my first male Pied Harrier. It lazily passed us by as if we did not exist. This is a beautiful bird, in its black, white and grey attire and easily rivalled the Black Harriers I had swooned over in South Africa. When serving in Burma during the second world war, the late Donald Watson, once described the male Pied as the most beautiful of the tribe, and how correct he was. Later that afternoon, while I was watching a large pool which contained several species of duck, a female Eastern Marsh Harrier appeared out of the blue, scattering the birds in different directions. At a glance, her plumage was similar in detail to an African Marsh Harrier, as she persisted in harassing the resident duck population, the majority of which were Common Teal. Hopefully, there would be more harrier days like this one at Kaziranga!

The next morning while making our way to the Eastern Range for the first time, I encountered a male Pied, hunting over a nearby rice field. This bird was within metres of several farmers who were working in the area and both appeared unconcerned by each other's presence. It quartered the area so gracefully, with wings held firmly in a dihedral V-shape and I was sorry to move off, even after 15 minutes of continuous viewing through binoculars. In the Park, I encountered two more males hunting simultaneously over a large marshy area, and observed one catching a frog, a short distance from my vantage point. Before leaving at noon, a male Hen Harrier was unexpectedly observed hunting in the same area, but it seemingly preferred the far bank of long grass containing small birds and unseen mammals. The previous evening after dinner, I had spoken briefly with Maan Barua (the owner's son). It was he who recommended that the Eastern Range was best for observing harriers. He was correct! Prior to leaving Northern Ireland, I had been in touch with Maan by email, as he is the official bird recorder for Kaziranga and Assam, and so his advice was most welcome.

Female Pied Harrier

Romy Ocon

The afternoon was again spent in the same general area, but only one male Pied and a female Eastern Marsh, were encountered, but a pair of noisy nesting Pallas's Fishing Eagles, were a welcome distraction, plus a flyby Crested Serpent Eagle. On our way out of the Park at 4.15pm, another male Pied was suddenly joined by a second individual, over the large rice field which we passed each morning and afternoon on our way to the Eastern Range. These two birds foraged without success for ten minutes, before splitting up and moving off in different directions.

On my return to Wild Grass, I received an invitation to have dinner with Maan's family, which was gratefully accepted, and there we discussed the possibility of observing all six harrier species at Kaziranga. Firstly, Maan told me that during the winter months it is possible, but with some difficulty, owing to the size of the Park (430sq km), to observe at least four harrier species. These being Hen, Western Marsh, Eastern Marsh and Pied. The male Eastern Marsh is also difficult to locate, with only five or six recorded here each winter. Secondly, there has only been one record of Montagu's Harrier in recent years, this being in 1996, and only a handful of Pallid Harrier records over the past decade (1994-2003). So I could categorically state, having heard it from the local expert, that it was highly unlikely, though not impossible, that I would be observing six species of harrier during my visit!

An early start on the 17th proved fruitful during my return to the Eastern Range. On this occasion I saw my first two female Pied Harriers hunting in tandem over the same marshy area that I regularly watched during my visits. Later in the morning a single male Pied, in dishevelled plumage, was observed and again he caught a frog, which are probably abundant in these wetland areas. For a change we went to the Western Range in the afternoon and from a man-made viewing facility began watching the large pool of water surrounded by long grass and juncus. At around 3.30pm, one, then two female Eastern Marsh Harriers appeared from an area to my left where small bushes and long grass were prominent. They immediately flushed the mixture of wildfowl and waders that were present. After several unsuccessful strikes by both birds they moved candidly away to try their luck elsewhere. All I needed to see now was a male Eastern Marsh, which was proving more difficult as the days quickly passed.

On the morning of 18th I was informed that a strike by the drivers, who take visiting tourists into the Park, would affect my itinerary. This was followed by a general strike the following day due to civil unrest by students in Guwahati. That morning, part of my package holiday included an Elephant ride, but sadly it had to be cancelled due to the drivers' strike. The morning was therefore spent in the adjacent grasslands and rice fields looking for foraging harriers, which was somewhat successful, with a male Pied giving me superb views at one location, followed by a female at another.

In the afternoon we had been given permission to visit the nearby Central Range, so Maan joined me in the field, as all the local students now had two days unexpected leave. He took me to an area a short distance inside the Park where he suspected a pair of Pied Harriers had bred in 2001. It was one of the few areas where the grass was much shorter

Immature male Pied Harrier

Mervin Quah

and ideal for breeding harriers, but we could not visit the probable nest site because of the dangers involved.

The National Highway was completely silent on the morning of 19th, as there was no heavy traffic that could usually be seen streaming eastward each day to Myanmar. Another morning was spent in the nearby fields, but no harriers were observed on this occasion. In the afternoon after an early lunch we were told it was now safe enough to drive to the Eastern Range and when we proceeded on to the Highway, we were, amazingly the only vehicle on the road at one stage. When we collected our armed guard at the entrance, he politely told me that he had been attacked and mauled by a tiger a year earlier and had luckily survived to tell the tale. He showed me several of the massive wounds that had been inflicted on him during this sudden and unexpected attack in this same area of the Park. Just like the Masai Mara in Kenya you do not go wandering around in Kaziranga National Park!

Just over three years later, I read in the Daily Mail in January 2006, that a Mahout was attacked by a Royal Bengal Tiger, when on board his Elephant in the Park. The attack happened as rangers tried to tranquilise the eight foot long, 350lb tigress who, with her two cubs had killed three cows from a nearby village. The idea was to relocate this huge cat elsewhere with her weaned cubs, but although the latter were sedated, the female escaped capture, with the Mahout suffering serious damage to his forearm, hand and the severing of two of his fingers with her bite. One of three pictures featured in the article, actually showed the tiger leaping approximately ten feet off the ground. My guard had a very lucky escape and to this day he still does not know how he survived!

On the 20th, we set off early for the Eastern Range and on the way to the Park, a single male Pied was again hunting over the rice fields to the right of the track. He suddenly plunged to the ground with his long legs fully extended, and I watched him depart with what was probably a rat, as a longish tail could be seen dangling freely as the bird flew away. When we reached the wetland area, I observed a male Eastern Marsh Harrier – the only one seen during my week-long visit. Its distinct plumage markings were evident in comparison to the female Eastern Marsh Harrier and Pied Harrier. The male's boldly patterned black and white body could be described as Pied, but the head, nape and breast were heavily streaked black, with white on the lower belly and legs also clearly observed. The back and upperwings were also black and spotty, but the secondaries and tail were grey, with his overall jizz, slightly resembling the male Reunion Harriers, I studied a year later in 2004. Its other name of Spotted-back Harrier is probably fully justified!

My afternoon watch was conducted from the viewing gallery in the Western Range. A female Eastern Marsh was present on a small bush for at least an hour, before creating more mayhem over the wetland. This time she was successful in ambushing a small unidentified wader species which she plucked in full view of my observation point. A distant male Pied, accompanied by a female were also seen but they did not appear at the wetland and flew away in an easterly direction, surprisingly, together.

The 21st was my last full day and so all of it was spent in the more productive Eastern

Range and the adjacent rice fields. Early morning appeared to be the best time to monitor harriers here, while afternoons were really good in the Western Range. During my week long visit I did not observe a harrier in the Central Range. The very tall Elephant Grass there is not ideal hunting habitat for them and the area is quite disturbed by numerous animals as well as people.

Almost every morning when I arrived at the marshy wetland lake, a family party of five otters, would entertain me before the arrival of the harriers, which usually appeared from a northerly direction, as they had to fly high to avoid a wooded area on the way in. This morning was no exception and before leaving for lunch, I was treated to male and female Pied Harriers. They spent 20 minutes quartering the area for prey, before flying south towards a grassy area, which we were to pass on the way out. Sadly, none were present when we left around 11.30am, but a male Pied was encountered once more as we drove slowly past the large rice field on our way back to the Wild Grass Resort near Kohora, for lunch.

A change of plan saw us arriving at the Western Range in the afternoon. We spent over two hours there but only observed a single female Eastern Marsh Harrier. In fact harrier numbers were by no means numerous over the week, probably 'conservative', would be an accurate description. I must add, that if we had been allowed to drive further north into the Park on the prescribed tracks, towards the banks of the Brahmaputra River, I may have observed more harriers but large areas of grassland were partially under water and sadly closed to the public. My week of observations revealed that both these species preferred to hunt over wet and marshy ground and very rarely in dry grassland areas, which is what one expects of the Pied Harrier.

On the 22nd I made the long journey back to Guwahati, and then my flight to New Delhi, via an unscheduled stop which gave me a fantastic view of the tall peaks of the Himalayas. A five-hour break at a hotel in Delhi gave me a welcome rest and time to reflect on this trip, before catching the 3.25am flight back to London and then on to my home on the outskirts of Belfast.

This was my fourth visit to India, in the past six years and I must say that undoubtedly Assam was the most attractive of all the locations I have visited. Because of the monsoon season it was so lush and green, and would not have been out of place in Northern Ireland. It was not easy observing and locating harriers at Kaziranga, but my efforts were finally rewarded and well worth the many hours of constant perseverance. My only regret was that I did not observe many male Eastern Marsh Harriers, which were probably not common here in any case. The only thing missing was not being able to find a notable winter roost site, which would have been a fitting conclusion to my visit. Anyway, there is always another day for that!

When my harrier colleague from South Africa, Rob Simmons, learned of my intended visit to Assam, he asked me to try and collect Pied Harrier feathers for DNA purposes, as he was still devoid of material for this species and the Papuan Harrier. Sadly, I was unable to oblige on this particular occasion, but in November 2006, my close friend

Roger Clarke became aware of our plight and donated several female Pied Harrier feathers (he had acquired from a friend) to Professor Michael Wink of Heidelberg University in Germany for research purposes. Unfortunately, these did not provide positive results, but others were forwarded in early January 2007 and I await the results which will hopefully reveal the vital phylogeny details of this species. Since then, I have managed to acquire more feathers from Roger's friend, Nigel Middleton. Nigel who is warden of Sculthorpe Moor Nature Reserve in Norfolk, diligently cares for a seven-year-old female Pied Harrier at his home. This bird was originally smuggled into the UK illegally by unscrupulous individuals.

The elusive male Eastern Marsh Harrier   *Lim Kim Chye*

The very rare male Reunion Harrier   *Laurent Brillard*

REUNION HARRIER AND MADAGASCAR MARSH HARRIER

Circus maillardi and Circus macrosceles

The **Reunion Harrier** was recently separated and elevated to a full species by Dr Michael Wink, (Simmons 2000), based on vital DNA evidence and morphological differences from its sister species, the **Madagascar Marsh Harrier.** The Reunion Harrier, or the Papangue, as it is affectionately known by the local inhabitants, is not only the rarest of all 16 harrier species, with a population estimated at less than 100 breeding pairs, but it also has the smallest range and is endemic to Reunion Island. Unbelievably, it is the only resident raptor on this 50 x 70km island and so I was not spoiled for choice, nor could I make the cardinal mistake of confusing this bird with another raptorial species, when viewing it for the first time.

In local folklore, the 'Papangue' (meaning harrier) or 'Maillardi Buzzard', is a protected and highly revered hawk-like bird, that starts life as a little brown bird and as it grows older, its plumage turns black and white (which obviously refers to the male), and more importantly, it is Reunion's only surviving bird of prey.

The Mascarene Islands, as Mauritius, Rodrigues and Reunion, are collectively known, lie in the Western Indian Ocean, approximately 700km east of Madagascar. Reunion is a volcanic island and is mostly mountainous and extremely steep. Three large and deep forested cirques (steep-sided semi-circular depressions), Mafate, Cilaos and Salazie, surround the highest peak (Piton des Neiges, 3,069m), and an active volcano (Piton de la Fournaise, 2,631m), in the southeast, is the second highest summit. The majority of the population live in the coastal lowlands, and the lower slopes of the island are totally cultivated, mainly with sugarcane. Intense agriculture and sprawling urbanisation, along with the construction of a dense road network, have dramatically affected the remaining natural habitat. Because of the height and inaccessibility of the higher parts of Reunion, many endemic species, such as the Reunion Harrier, have survived in much healthier numbers, than on neighbouring, Mauritius and Rodrigues. Therefore, the distribution of the Reunion Harrier was mostly concentrated on steep slopes in heavily forested areas at mid-elevations (300-700m), and occasionally down to sea level, but rarely above 1,200m. Consequently, foraging was not as diverse as first thought but this harrier was more widely distributed in areas where native and degraded woodland existed on steep inclines and amongst habitat comprising montane vegetation, cultivated (sugarcane) fields and pastures, open grasslands, savannas and occasionally coastal.

On the 29th October 2004, I flew to Reunion Island, via Belfast, London and Paris, for one full week. The second week would be spent on the world's fourth largest island, Madagascar. On arrival at Roland Garros Airport, on the morning of the 30th, I soon discovered that Reunion was one of four overseas departments owned by France and that the local currency was the Euro! After taking charge of my hired car, I set off for the picturesque village of Hell-Bourg in the Cirque de Salazie, where I would be based for the next four days. This area was well renowned for regular sightings of this species, so I was hopeful that I would encounter one or two birds before arriving at my accommodation.

Adult male Madagascar Marsh Harrier

Lily-Arison Rene de Roland (c/o Peregrine Fund)

When I left the coastal highway at St Andre and made my way inland towards Salazie, the road became windy and narrow as it ascended steeply into the hills which were heavily forested on both sides. The greenery, which in certain areas was interrupted by cascading waterfalls that ran into the River du Mat, was spectacular to say the least. I could not believe that this was the habitat in which I would be observing Reunion Harriers for the remainder of the week.

My first Reunion Harrier, a male, was observed at 11.00am as it flew from trees close to the road. It was carrying what I considered to be a rat and made its way across the river, where it disappeared into a stand of deciduous trees that were lining the hillside. On parking my car, I observed a second male, circling high above the trees and it also disappeared in a similar manner to the first. I waited for another 20 minutes hoping to see an adult female, but none materialised on this occasion. Driving slowly, mainly due to the many dangerous bends and steep hills, I eventually reached Hell-Bourg for lunchtime.

At 2.00pm, I commenced my studies of this strange harrier south of Hell-Bourg, where I firstly had superb views of a foraging male which appeared close to a small house and its surrounding garden. Then my first female appeared from nearby trees which immediately interacted with the male. This attractive but smallish male, in his black, grey and white plumage was superb to watch as he effortlessly weaved his way through the foliage of the nearby trees, with the dark brown female and her conspicuous white rump patch, or uppertail coverts clearly visible, even to the naked eye.

The following morning I was on the road again, only this time I had rejoined the coastal highway to explore habitat around the periphery of the island. This isolated species had been described by Bretagnolle *et al* 2000, as the Reunion Marsh Harrier, so now I wanted to find the wetland habitat associated with true marsh harriers, and hopefully the birds themselves. I drove as far south as Ste-Rose, driving inland at St Benoit and Ste-Anne, but by mid-afternoon I had failed to discover any suitable habitat. When I returned via St Andre, on my way back to Salazie and Hell-Bourg, I observed several Reunion Harriers, and where were they? – in the hills soaring high above the forest canopy. The remainder of the afternoon was spent in this area and I managed to encounter at least eight birds, five of which were males, before heavy rain brought my studies to a premature end.

The next day, 1st November, was spent at known harrier vantage points south of Hell-Bourg and east to St Andre, but my observations were cut short on several occasions due to heavy rain and mist that descended on the surrounding hills. Close to where I had observed my first Reunion Harrier, I encountered two males hunting in tandem over a wooded hillside close to the main road. After watching their hunting technique for a full 10 minutes, I discovered that one of them was disappearing into the trees and then suddenly appearing again below the branches. This bird was actually flushing a small flock of birds from the tree-tops where they had obviously been perching and during this unfamiliar activity it finally appeared again with one firmly clasped in its talons. The second male by this stage had flown across the road and began repeating the same behaviour a short distance from where I was standing.

A diligent pair of nesting Reunion Harriers   *Laurent Brillard*

A few minutes later there was a sudden commotion in one particular tree and, as I anxiously watched the outcome, a small party of bats emerged from the tree, with the harrier having caught one of them. This bird had actually flown into the foliage and captured one of the bats that was roosting below one of the many branches. 'What strange behaviour for a so-called marsh harrier', I muttered quietly to myself.

My fourth and final day in this area (2nd) was mainly spent on the coastal highway and inland locations from St Andre to the capital, St Denis, in the north of the island. Once again, I found no areas with suitable marshy ground containing reeds that are a necessary requirement for marsh harriers. Slightly disappointed, I returned along my familiar route towards my accommodation in Hell-Bourg, taking in an occasional harrier soaring high in the hills like an eagle or goshawk, and also the breath-taking scenery in this particular cirque (Salazie).

After breakfast the next morning, I eventually rejoined the highway and drove southeast to St Benoit, where I turned inland towards my new accommodation in Confiance. This area was at a much lower altitude than the previous location, close to the coast and in an agricultural area where sugarcane, was seemingly prominent everywhere. When I checked in later in the afternoon, my hosts asked me what the purpose of my visit was to Reunion, and when I said I had come all this way to study the local harrier population, they smiled and quietly retorted – 'the Papangue'! This lady knew her birds, and she told me that if I drove a short distance to the main road I would probably see this bird hunting in open areas where the sugarcane had been recently harvested. Before dinner I watched the area without success, but apparently she had meant first thing in the morning after breakfast. I was nevertheless slightly pessimistic.

On the 4th, after breakfast, I drove the short distance to where I had sat the previous evening and around 9.30am a male suddenly appeared from the misty hills to my right. He came right into the field that was partially harvested and began foraging for prey and then briefly landed on a small bush to my left. Hunting commenced again and on this occasion he caught what I took to be a small bird. Minutes later a second male arrived in the same area, but he left when the original male returned unexpectedly. Around 10.00am, a welcome female arrived giving me superb views of her chocolate brown plumage and the slight traces of buff on her seemingly short upperwings. The male suddenly joined her close to the road and I noticed that his wings were also short and rounded, with both sexes more akin to a goshawk, than the familiar long wings of a harrier. This was the closest I ever got to the Reunion Harrier during my week-long stay on this island and it was also the best location to photograph this bird. All my previous attempts were from a distance, but at least I had a half-decent record of this up to now, elusive harrier. Sadly, two tractors came from a nearby laneway laden down with sugarcane and both birds dispersed towards the sloping hills in the distance. The remainder of the day was spent in the nearby hills, watching several males and females in – La Plaine des Palmistes area. When I arrived back at the location where I had watched earlier in the morning, a distant male was observed for several minutes, but then he suddenly disappeared towards the coastline and was not seen again. This was my final sighting of a Reunion Harrier.

It is important to add, that the Reunion Harrier was previously described as a member of the marsh harrier complex, which consisted of six, probably seven species if one includes the little known Papuan Harrier. Having intensively studied this species, I can conclude that I found no evidence of these birds inhabiting marshes or wetland areas. To the contrary, they preferred higher ground where densely wooded hillsides existed. Perhaps in the past this species truly was a marsh harrier but has now adapted to a changing environment on Reunion Island. I personally doubt, therefore, whether the Reunion Harrier really belongs to this elite group of harriers.

There are only three main marshes on Reunion, all of which are located close to the sea, with several smaller ponds and wetlands further inland (L Brillard pers comm). Regrettably, problems exist at these important sites due to indirect disturbance from humans and encroachment from developers, and this has probably been responsible for forcing many of these rare birds into the nearby hills and cirques. In these magnificent cirques, the Reunion Harrier is described as uncommon in Cilaos, extremely rare but mainly foraging in Mafate and as I duly witnessed, common in Salazie.

The aerial behaviour of these birds, particularly in the Cirque de Salazie, was so reminiscent of the County Antrim Hen Harriers, as they also spent much of their day gliding and loafing in tree-lined habitat. This led me to speculate at the time, whether these Reunion Harriers could also be tree-nesting? Then, unexpectedly I received from Laurent Brillard , a photograph of a pair of Reunion Harriers nesting on the top of the tropical vegetation in the steep-sided Plaine des Cafres and not on the ground, as is generally the case with this species! The nest was almost 1,000m up, close to the summit and not near their supposedly traditional breeding grounds in low-lying wetland areas. Laurent explained, that the ground in these hilly areas of Reunion sometimes means the nest is located on a small bush on the densely vegetated hillsides, though the practice is not thought to be common.

Owing to the previously mentioned pressures in the lowlands, these birds have seemingly adapted to an unfamiliar form of nesting and I think it is only a matter of time before a nest is found in a tree. I suspect this may already be occurring and remains undiscovered for the present. I would, therefore relish another visit to Reunion, only this time during the peak of this species' breeding season!

The next morning (5th), I arrived at the airport for a noon flight to Antananarivo (known locally as Tana), in Madagascar. The island of Madagascar, nicknamed 'the 8th Continent' because of its diversity of species, lies some 400km off the east coast of Africa, and is legendary for the profusion of its wildlife and flora, 80% of which is found nowhere else on earth. It is part of the group of islands called the Comoros Islands. Madagascar, is probably more famous for its 33 species of lemur (found to date), than its birds of prey, but several rare and endemic species of raptor also exist on this amazing island, including the Madagascar Marsh Harrier, my target species.

Ornithological historians also tell us that Madagascar was formally home to the largest flightless bird that ever lived – the Elephant Bird.

Until recently (2000), the Madagascar (or Malagasy) Marsh Harrier was considered conspecific with the Reunion Harrier, but after a re-analysis and DNA tests, they have now been declared two separate species. As I had now observed and studied good numbers of Reunion Harriers, it would be interesting to see what morphological differences exist between these two species and hopefully these would be revealed to me. The main population is found on Madagascar itself, but birds are also found on the other four main islands in the Comoros group. The population status and distribution is not well known and Langrand (1990) only highlights nine areas on this huge island where this species may be found breeding. Therefore, it is regarded as an uncommon breeding bird and is now infrequently observed coursing over marshes, fallow rice fields and grasslands. As a result, Birdlife International (2000) has classified the Madagascar Marsh Harrier as globally vulnerable!

On arrival in Tana I was met by my guide for the next five days, Jocelyne Randimbison and our driver. Within 45 minutes of my arrival, we were on our way in a 4x4 jeep, to Ambohitantely Special Reserve, which is approximately 140km northwest of the capital Antananarivo, in the central highlands of Madagascar. Ambohitantely is an extremely large area of mainly degraded grassland and primitive forest covering some 5,600ha, at an altitude of 1,300-1,600m. Unbelievably, this is where the Madagascar Marsh Harrier is found breeding. I would be camping here as there were no hotels in the immediate vicinity of this reserve. Arriving late in the afternoon there was not much time for harrier studies, due to fading light, tents to be erected, a fire to be lit and dinner to be cooked hopefully, by Jocelyne.

Prior to leaving Northern Ireland I had been in regular contact with Russell Thorstrom (from the Peregrine Fund, USA), who had studied this species during the 1997 and 1998 breeding seasons at Ambohitantely, and who still works on the island, today. During his studies, he had managed to locate 11 nests, of which eight nesting attempts were documented over the two seasons. This was the first ever breeding study undertaken of this species and it sadly showed that this harrier reproduces at a relatively low rate, due to constant disturbance and persecution from humans.

After an uncomfortable night, followed by a hearty breakfast and an early start the next morning (6th), we began driving slowly around the Reserve in the hope that we would locate a harrier. As it was the breeding season we were more likely to encounter a foraging male rather than a female. At around 9.00am, a male suddenly appeared over the grassland to my right, carrying prey – a small bird, presumably for an incubating female. When he suddenly disappeared behind a line of tall trees and down a steep hillside, I jumped out of the vehicle and ran quickly to this location. He was now circling and calling excitedly and peering down into a low-lying marshy valley – was this a nest site? I was not to be disappointed, for within minutes the resident female had taken to the air and a successful food-pass was completed before my eyes.

A remote mountain village in the Cirque de Salazie, Reunion Island – typical harrier habitat

Don Scott

Both the male and female were extremely large birds, and immediately noticeable was their exceptionally long wings and prominent tails. The male was similar in plumage detail to the male Reunion Harriers I had studied the previous week, but he was immense in comparison. Similarly, the female was even larger and paler-plumaged than her sister species on Reunion. Her body size alone made her one of the largest harriers I have ever observed. This species truly was a marsh harrier, unlike its counterpart on Reunion, which appeared more specialised and more dependent on tree-lined areas than marshes. Within minutes she had consumed the prey item and returned to her nest near the centre of the marsh as the male obligingly circled high above me.

We drove deeper into the reserve to find more nesting harriers. The habitat was similar to that we had left earlier. At around 11.30am, another male was observed carrying prey and as we slowly followed him; he too dropped down into marshy habitat. As he called out and briefly circled low above the nest, the female arose immediately and received the prey, which appeared to be a large frog. On this occasion the female took the item back to the nest. Had she chicks, or was she in the early stages of incubation? The male surprisingly remained close to the nest and began preening for a full ten minutes before disappearing away to my left. Further observations from two different locations later that afternoon revealed two more foraging males, but no more nest sites were discovered.

During the night, heavy rain began beating on my tent followed by strong gusts of wind, which is probably normal for this area at such a high altitude. When morning dawned (7th), it was bright and sunny, although it did remain chilly until mid-morning. We travelled back to the first site we had found the previous day and awaited the arrival of the male. He arrived just before noon, carrying what we thought was a very large rat-like mammal. As the female flew up to greet him a short distance from us, I could clearly see that the item he had given her was indeed a large rat. Feeding immediately commenced in the right-hand corner of the marsh (her usual area for taking prey to) and lasted for 15 minutes. She did not consume it all and carried the remainder back to the nest, strangely, in her beak. The male flew extremely close to where I sat on the sloping bank and I actually managed to get a few decent photographs of him before he wandered off elsewhere.

In the afternoon I visited the remnants of the nearby forest, and managed to get superb views of a soaring Madagascar Harrier-Hawk *radiatus sp.*, a small party of Brown Lemurs, several species of chameleon and, rare orchids growing on the rotting and lichen-covered branches of several fallen trees. I was very excited at seeing this species of harrier-hawk, having studied the diet of its sister species *typus sp.*, in The Gambia, West Africa, years earlier.

The next morning (8th), after an early breakfast, I arrived at the first nest site just before 9.00am, and had taken up my position on top of the sloping bank, to await prey deliveries. From that time until noon, five items were brought to the female, comprising three birds, a rat and a large fat frog. Deliveries were noted between 9.30am and 9.50am, 10.20am, 10.55am and finally at 11.05am. The male for some reason always arrived at this site from an easterly direction. When he had not reappeared by 11.30am and with the female also absent, I paid a brief visit to the nest with one of the local guides. I had to wade through deep marshy ground, over my knees in places, before I reached the nest site. The nest was set amongst juncus, grass and sedges and was well above water level. It contained three largish blue-green eggs, which were hastily photographed for record purposes, before leaving. As we retreated, the female (she was huge), appeared high above us, calling in an agitated manner, and so I managed to photograph her as well.

We now moved to the second nest site further north to await prey deliveries there. These were recorded from 1.00pm until 5.00pm. During these four hours, the male consistently brought an item to the female at hourly intervals. These were a small lizard, one large and one small bird and then a small mammal, all of which were unidentified. After the final prey delivery, the female, was observed mobbing a Madagascar Buzzard which accidentally strayed into her territory. This site was slightly drier under foot and more easily reached than the previous one. It too contained three eggs which I photographed before we left. When I arrived back at base camp around 6.30pm, my legs were extremely sore after wading through the mire on two occasions, but Jocelyne had prepared an excellent chicken dinner and the pains soon disappeared.

The 9th was my final day at Ambohitantely and before leaving at around 3.00pm, I briefly visited both nest sites. During the course of my visit I had managed to locate at least four breeding pairs in the surrounding hills and valleys, and at least two unattached males. The third nest site was finally found a short distance from the second site and it was briefly watched between 10.30am and 2.00pm. In three and a half hours, only two deliveries were made, one at 11.18am and another at 12.20pm. Two Pied Crows were ever present at this site and they continually harassed the male or the female when they appeared, before returning to perch in a tall pine-like tree.

On our way back for a quick cup of coffee and to gather up our possessions, I observed a young male in eclipse plumage (my first sighting of one), and then a few kilometres from camp, a single immature female was seen pouncing on an unseen prey item. When she slowly took to the air, I could see that it was a thin brown snake about 1 metre long, which the harrier had seized with both talons. As she flew off to our right you could clearly see the snake trying unsuccessfully to squirm from her grasp. This was also a first for me as I had never observed a harrier of any species catching a snake before. I must add that during the day in Ambohitantely, there was always some sort of harrier activity observed, but mainly of adult males, as they hunted diligently over the vast grassland and sloping hillsides for prey.

I noticed that the ground in this area was very porous with the hills showing signs of erosion in many places, and it was probably this that formed the low-lying and almost circular marshy valleys that these harriers were utilising. It seems likely, therefore, that the Madagascar Marsh Harrier may have nested in much lower lying areas in the past and like the Reunion Harrier, was forced through loss of natural habitat, to occupy high altitude territory. This habitat would have been more suitable for nesting Hen Harriers and

Top and Bottom: Male and female Madagascar Marsh Harrier   *Lily-Arison Rene de Roland (c/o Peregrine Fund)*

I know of no other marsh harrier species, nesting at such high altitude. One serious worry I have about this area is the indiscriminate burning which occurs on the periphery of the Reserve on a daily basis. Trees, are continually burned for charcoal, and sold as you enter and leave Ambohitantely. Sadly, they are never replaced and repeated grass-fires and illegal timber harvesting is quite common in this area and in other parts of Madagascar, where other vulnerable species are at risk of extinction, such as the irresistible lemurs and two-thirds of the world's chameleon species! There are three species of lemur resident in Ambohitantely. I only observed one however – a small troupe of Brown Lemurs foraging close to where several trees had been felled, sadly within the Reserve itself.

On a more relaxed note, I did learn a little bit of the Malagasy language as when I identified a male Madagascar Marsh Harrier, it was referred to as *Lahy* , and the female as *Vavy*. The species itself is known as *Fanindry* (which flattens), and obviously refers to the harrier's disk-like facial features.

On our way back to Antananarivo we stopped to admire a male, which we observed harrying a small bird over a paddy-field, south of Ankazobe. This was the only harrier, we observed on the way to Tana, and surprisingly none were encountered when we travelled up to Ambohitantely, on the 5th.

After a shower and a good night's sleep in a comfortable bed I was collected the next morning (10th), by Jocelyne and her driver and taken to the airport around 11.00am, as my flight back to Reunion was leaving at 1.20pm. On arrival in Reunion, I collected my hire car at the airport and proceeded to my accommodation in Confiance, where I would be spending my final evening. My last day (11th), was spent watching the occasional Reunion Harrier, but since the breeding season does not commence here until late November (seemingly much later further east), I was unable to witness any early nesting behaviour which was most disappointing. I did however observe (from a distance), what appeared to be a bird in juvenile plumage that had fledged the previous year which flew from the trees on the arrival of both adults? After a reasonable and final day in the field, I made my way back to the airport, for my three long flights back to Northern Ireland.

Finally, I must mention the threats, to this species and its habitat! Threats to the Madagascar Harrier are now numerous and include dry season grassland fires that usually occur during the nesting season, loss of vital marsh and grassland habitat and blatant human persecution to apparently protect poultry in local villages. Therefore, in Ambohitantely, and most likely throughout Madagascar, there are mainly three major human threats towards this species. Firstly, adults birds are persecuted for food and because they pose a threat to domestic fowl. In one horrific case in 1996, Ignace Randriamanga (2000), met a poacher who displayed the carcasses of 13 Madagascar Marsh Harriers he had killed. Local people also consume the eggs and nestlings as a source of protein. Secondly, during every dry season (April to October), the grasslands of Madagascar, especially the high plateau region, which includes the Reserve, are burned by humans to stimulate (green) growth for cattle fodder and also for land clearance. As a result, several harrier nests are known to be destroyed during the vital incubation period, by uncontrolled fires. Thirdly, the conversion of low-lying marshes and wetlands to ricefields for human food production seriously reduces vital nesting habitat for this magnificent species.

The current vulnerable status, therefore, awarded to the Madagascar Marsh Harrier which is based on the threats to its habitat and its sparse distribution throughout Madagascar, is fully justified. Although it has been recorded at a number of protected areas and National Parks, these Reserves have been mainly established to preserve forested habitat and sadly have limited grassland and wetland protection. Biologists in particular, need more information on the Madagascar Marsh Harrier's population size and breeding ecology, in order to provide vital and long-term conservation strategies and full protection, in the near future.

Summary: *A week or so later, I sent the following roughly drafted email of my findings on Reunion and Madagascar to my harrier colleague in South Africa, Rob Simmons:*

"I have recently returned from the above locations after a week of harrier observations on each island. The Reunion Harrier is most definitely a specialised harrier species as it seemed to spend most of its time hunting and soaring at altitude in the cirques. I was staying for the first part of the week at Hell-Bourg, which was around 950m above sea level. On occasions, these birds were even higher as they foraged amongst the heavily wooded and vegetated hillsides. On at least two occasions, I observed them flushing small bats, presumably from under the branches of trees, and also hunting small birds from the tree-tops, in similar fashion. One thing was very noticeable about their wings they appeared shorter and rounder, which is unusual for harriers and they were in habitat that was more akin to buzzards, goshawks and to a lesser extent eagles. Obviously the harriers' shorter wings are adapted to allow them to hunt amongst trees and dense foliage. In many ways their habits resembled that of the tree-nesting Hen Harriers we have here in Northern Ireland, especially when young birds flew from the trees begging for food from the arriving adults. If ever there was a case for tree-nesting, then the behaviour of these birds makes them prime candidates. I searched the lower climes of the island, for marshy and wetland areas suitable for nesting, but unfortunately without success. The lower areas are now heavily populated and intensely covered with crops, mainly sugarcane. I did though notice, that in areas which had been cleared of sugarcane the birds would forage over them to a certain degree, but they seemed to prefer the wooded areas, which were also nearby.

"In comparison to the Madagascar Marsh Harrier, there is no way, based on my observations, that the Reunion Harrier can be classed as a marsh harrier!

"At an area in Madagascar (Ambohitantely), I managed to locate at least four breeding pairs of Madagascar Marsh Harrier with an immature male and female also present. The latter was observed catching a snake in 'Secretarybird' fashion. The Madagascar Marsh Harrier is a most impressive and extremely large harrier. The female and male of this species are larger than the Reunion species. Females are a much lighter brown, and the males are not as dark on the upperparts as the Reunion. The Reunion females and immature birds are more chocolate brown on the head and upperparts, and the males are a much darker black, particularly on the head, neck and upperwings. One of the Madagascar Marsh Harrier nests I visited, was set amongst rushes *juncus sp* and sedges, and in much more treacherous ground than the Black Harriers I had studied in South Africa. The three eggs were similar in colour to other harrier eggs.

"Of the two species, the Madagascar Marsh Harrier was easier to see, because it foraged and nested at much lower levels than the Reunion Harrier, apart from some that were found high up in hilly grassland areas approximately 1,500m asl. Apparently, the numbers of Madagascar Marsh Harrier are dwindling, but the Reunion birds, appear to be plentiful (to a certain extent), as they have seemingly been pushed up into the hills. The Reunion Harrier is lucky in the sense that there are no other large predatory birds on the island and so it can monopolise prey at both high and ground levels."

Rob's reply several days later read: "Finally, had a chance to sit down and consume your very interesting details on the differences between the Reunion and Madagascar Harriers. I agree the Reunion bird is a prime candidate for a tree-nesting bird and I'm surprised that no tree nests have been found. Perhaps like your Northern Ireland birds they are difficult to locate unless you have a good vantage point overlooking the canopy. Amazing, that they are so different and no one bothered to describe them as separate species!"

J-M Thiollay has written a paper on them, which appeared in *Raptors at Risk* (2000), edited by Meyburg and Chancellor for WWGBP. They reckoned the breeding population was less than 100 pairs, concentrated at mid-elevation in forested areas. Since the island, was completely forested when the first British sailors described it, the harriers surely must have nested in trees then. Shame that modern man got there before the biologists!

Thiollay *et al* reckon it is now rarer than the Mauritius Kestrel, once the world's rarest raptor – a sobering thought!

I have seen the Madagascar Marsh Harrier and yes it is a big bird. Some of the island's inhabitants are gradually destroying vital and much-needed habitat. I think within 25 years, the harriers will only be able to survive on the specific reserves on the island – another sobering thought!

Footnote

Most recent correspondence with Russell Thorstrom in November 2006, now describes the Madagascar Marsh Harrier as critically endangered, as surveys in 2005 to 2006, only found around 77 pairs. Thankfully, this species' status in the country will immediately be elevated from globally vulnerable to endangered and hopefully saved from extinction! During the harrier survey the Peregrine Fund team re-discovered the Madagascar Pochard – a bird thought to be extinct since 1991 – so when other remote areas of this huge island are searched in the future, they may reveal additional numbers of Madagascar Harriers!

Madagascar Marsh Harrier nest

Don Scott

Adult male Cinereous Harrier   *Josh Larsen*

CINEREOUS HARRIER AND LONG-WINGED HARRIER
Circus cinereus and Circus buffoni

The **Cinereous Harrier** is found breeding in southern Brazil, through most of Argentina to Tierra del Fuego in the south, being most common in the Patagonian Steppes. Its range also includes Colombia, Bolivia, Paraguay, Uruguay and central and southern Chile in the west. It prefers to nest in mainly grassy and scrubby areas close to hollows where the vegetation is denser and also in montane regions where moorland exists. Nests are also found in wet ground or in damp grass amongst rushes *juncus sp* and other local vegetation. The Cinereous Harrier occurs furthest south of all harrier species and has even been recorded on the Falkland Islands at the southern tip of South America. The Falkland Islands' population, which is now virtually extinct, was formerly recognised as a separate race – *Circus cinereus histronicus*.

The **Long-winged Harrier** is found breeding in Colombia, Venezuela, Guiana, Trinidad and probably Suriname, south Brazil, Bolivia, Paraguay, Uruguay, central Argentina and possibly Chile. It is the only species of harrier that breeds north and south of the equator and is regarded as the largest harrier, though by no means the heaviest. It favours extensive marshes with tall reeds for nesting and hunting, small lagoons, rice fields, wet grasslands and even wet ditches close to main roads are regularly frequented in search of food. They rarely hunt over dry fields, but simply pass over them on their way to other wetland locations.

Just three weeks after arriving home from Madagascar and Reunion, I was on my way to Argentina in South America, to study breeding Cinereous and Long-winged Harriers. A long overnight flight was anticipated to Buenos Aires via Sao Paolo in Brazil, but to my surprise I arrived fresh and well when I was met by my guide Sergio Corbet, on the morning of 5th December 2004. A pre-arranged and typical fast taxi ride from the international airport to the domestic airport saw us eventually depart for Rio Gallegos, in southern Patagonia. This flight took around three hours and when we arrived early in the evening we were kindly received by our hosts Ludie and Nancy Henning. At the time I was completely unaware that I had an extra long car journey ahead of me – it was actually over 450km to my destination at Puerto San Julian! Argentina, is the second largest country in South America and with a length of over 2,000 miles, this journey was probably considered by locals as just a short jaunt up the road.

I soon got my first views of the wide open and seemingly desolate and barren plains that were Patagonia and every few kilometres Guanacos and small numbers of Lesser Rhea would appear to relieve the boredom. The high and cooling winds coming from Chile and the high Andes to the west had obviously stripped the land of its vegetation and trees and I found it hard to believe that this was where I was coming to study breeding Cinereous Harriers! After a short break for a welcome evening meal, Sergio and I eventually arrived in San Julian at around 1.30am – and if I was not complaining about being tired earlier in the day, I was now completely exhausted.

It is wonderful what a good night's sleep can do, for after a shower and a hearty breakfast we were soon on our way to Punta Tumba, where the Cinereous Harriers were

The appropriately named, male Long-winged Harrier   *Josh Larsen*

apparently breeding. Birds had first been observed in this area on 24th September by Ludie and Nancy and also photographed by their friend Natalie Collm. Territorial displays, sky-dancing and food-passes were observed on 1st and 3rd October, with at least six birds seen by them on that occasion. Seven further visits to the area from 6th October to at least 8th November, saw similar behaviour but copulation was noted between two pairs on the latter date, with disputes between males and resident Chimango Caracaras, being an almost daily occurrence. Chimango Caracaras are Argentina's most common bird of prey in towns, cities and the wider countryside. Interaction between both species was not good for breeding harriers, particularly as the Caracaras appeared to rule the roost in this area, as did a pair of Great Horned Owls. On 3rd December, a few days prior to my arrival in Argentina, a female Cinereous Harrier, had been observed feeding on an Elegant Crested Tinamou chick, another species that breeds in this area. My informants rightly pointed out, that these chicks were the favoured prey of these harriers. One brood of at least ten Tinamou chicks had now decreased to only two or three by the aforementioned date. Anyway, I had five full days here to observe behaviour and as many birds as possible and to find at least one nest site for scientific purposes.

As I initially thought, the weather was quite warm, as we walked about this desolate area in search of breeding harriers. However, as the morning progressed the so-called never-ending wind which inspired Charles Darwin to call Patagonia – The Whispering Land, soon made us feel uncomfortable. As we walked by the tallest and steepest hill, I immediately christened it 'Mount Harrier', for it was here I observed my first male and female Cinereous Harriers at around 11.30am that morning. By mid afternoon, there were at least four males and three females hunting in this general area, but no food-passes or early signs of nesting were noted. When we returned for an evening visit with Ludie, there were even more males present – at least eight – but seemingly very few females, which Sergio had already noticed. Hopefully, our observations tomorrow would be more positive and conclusive!

The next morning, the 7th, we decided to reach the area by 8.00am hopefully to observe food-passes and pinpoint a potential nest site. The 'Mount Harrier' area was again a hive of activity, with up to four males just milling and seemingly seeking early morning thermals to spiral over and around the hill, which we duly climbed to get an all-round view of the surrounding area. In the distance a male could be seen hunting over a small island across the bay and a short time later we observed him returning with prey. We unfortunately lost sight of him at a critical moment and totally missed the crucial food-pass to the female and her eventual landing location. Later that morning, I observed what was probably the same male hunting once again over the lush green island and then returning with what we both agreed was a small rat. Being about 250m away and having to walk over the brow of a small hill we again missed the food-pass. The female presumably returned to the nest immediately? Our luck was in for once as the male proceeded to circle low over the nest for several minutes before dispersing, in similar fashion to the behaviour noted at Hen Harrier ground nests in the past.

Adult female Cinereous Harrier

Natalie Collm

We now began walking slowly along both sides of a bushy ravine at the bottom of two small hills where the undergrowth had been replenished by occasional rain in this arid environment. I decided to walk halfway along the ravine and a female suddenly arose from almost under my feet. There on the ground and completely concealed in the bushes and long grass were three downy chicks approximately 7-10 days old. There was virtually no nest as such, although part of the surrounding grass had been compressed by the female and her chicks, to give an impression of one. The small bushes and surrounding grasses were at least 1 metre high where the nest was located and probably half a metre high elsewhere along the ravine. I must admit that the nest was well hidden from predators. A fox and a skunk had been seen earlier in the morning, but thankfully not near the nest site. I was excited by this find and photographed the nest for record purposes, before retreating from the area. Around noon we both watched a spectacular food-pass, from a nearby hill which overlooked the area and this seemed to impress Sergio, as I heard him utter – what a sight!

Three young Cinereous Harrier chicks   *Don Scott*

We decided to return early in the evening and watch the area until sunset as quite a few unattached males were present during our previous visit. Our watch was briefly interrupted by a pair of elegant Aplomado Falcons, as they swiftly passed by us calling in an agitated manner. We observed at least four male harriers without partners prior to sunset, with at least two of them sky-dancing to attract a mate, but without success. Three of them then pre-roosted on the surrounding bushes and as it came close to sunset they all dropped to the bare sandy ground close to the road to roost for the night. The fourth

Male Cinereous Harriers

Natalie Collm

individual had flown elsewhere to roost and did not join the three other birds. There was very little ground cover even for nesting pairs and so the birds probably felt safer on the open ground. In fact, I had remarked to Sergio on several occasions, that this was the worst breeding harrier habitat I had ever witnessed, even surpassing that of Hen Harriers in County Antrim.

The next morning we were all invited on a boat trip by Natalie, who is a local nature guide in San Julian. She took us to a large colony of nesting Magellanic Penguins and serene looking Blue-eyed Cormorants, plus a host of other seabirds. Spectacular attacks on several young and isolated penguin chicks were made by a Northern Giant Petrel, which tried not only to catch them, but also fly off with them. On the return journey our boat was escorted back to shore by a friendly pair of black and white Commersons Dolphins, which the pilot of the boat took to the water with for at least 20 minutes.

Our harrier activities resumed later in the afternoon and we decided that we should remain in the area until just before sunset. Once again harrier activity was slow to materialise and only a passing Short–eared Owl, relieved the boredom early in the evening. Then just before 7.00pm, two male Cinereous Harriers, ghosted into view followed by three more, minutes later. Once again nearby bushes were utilised, but not before two of the birds performed brief sky-dances, in full view of the three pre-roosting birds. As they passed before us they could be clearly heard calling in an excited manner, but again their efforts attracted no wandering or single females. Similar to the previous evening, three males roosted on the bare ground, one of which I managed to photograph in the failing light. The two other individuals flew off in a westerly direction.

The morning of the 9th saw us split up and go off in different directions to hopefully locate other breeding pairs in this vast scrubland, which was easier said than done. Walking in the area was reasonably easy as none of the vegetation was more than ankle height in most of the places we visited and it was only in the low-lying ravines or gullies that we encountered breeding harriers. When we met up at 1.00pm, we reckoned that four pairs were definitely breeding, and that there were another five unattached males, giving me an estimated total of 12-16 harriers observed daily. I believe that the four or five wandering males, which returned here each evening did so as they knew other birds, would be present and possibly an unattached female. The old adage of 'safety in numbers' appeared to apply to this breeding area. As we had no personal transport, it was difficult to explore other areas and we relied on being left in Punta Tumba each morning and evening which was probably no more than 8km from San Julian, at the most. As we would probably not encounter any more breeding birds between here and town, we decided to call it a day around 2.30pm and walk slowly back to our destination.

The 10th was our last full day here and so we decided to go over old ground until at least lunchtime. We sat down on the hill that overlooked the nest I had found earlier in the week to check that all was well. Just before 10.00am, the male duly arrived with prey in his talons and began circling and calling above the nest. The female without warning arose and joined her mate a short distance to the right of the nest and a successful, but on this occasion, unspectacular food delivery, was completed in mid-air. In the space of three minutes she had returned to the nest carrying what I believe was a small rodent-like creature. The male continued to circle low above the nest for several minutes before flying off towards the island where he tended to do the majority of his foraging. More prey was delivered to the nest at 11.30am which we got to observe before Nancy arrived to deliver a picnic lunch, prior to leaving for Rio Gallegos to meet her brother who was arriving from the USA. Our afternoon session eventually concluded around 5.00pm with really nothing further to report, and Ludie arrived to transport us back to San Julian.

Our taxi arrived the next morning at 9.00am to take us back to Rio Gallegos Airport, as we had an early afternoon flight to Buenos Aires. At least it was bright and sunny the whole way back and so I could keep an interested eye out for other Cinereous Harriers that may be in the area. Thankfully, our driver drove at a respectable pace and about 20km from the airport a male was seen hunting over lush green terrain to our right. Further along this main road and around 16km from our destination we had to slow down due to a police road check and there to our right over similar habitat were two males hunting almost side by side. This was a very damp area hence the lush green grass and in total contrast to the arid habitat at Punta Tumba. In areas like this, there was probably more prey to be had in the form of small birds and mammals which appeared to be uncommon further north.

On arrival in Buenos Aires, and as our flight had been delayed on the way back, our priority was to pick up our hired car before the office closed. After completing the usual formalities and a rather speedy guided tour of the huge city of Buenos Aires, we eventually made it to Route 9 of the Pan-American Highway, for our two-hour journey north to our hotel in Campana. I was now on my way to monitor the aptly named Long-winged Harrier in the legendary Pampas area of Argentina!

The vast and rich grassy plains of the Pampas were originally Gaucho (cowboy) country and cover almost 25% or 750,000sq km of Argentina, extending into neighbouring Uruguay and west towards the Andes. The western part is dry and largely barren but the humid eastern section where I was going is the nation's economic heartland where large herds of cattle and horses can be seen grazing in lush and wet green fields that are more akin to Northern Ireland than Argentina.

After a good night's rest and an early breakfast our first port of call on the morning of 12th saw us arrive at the Reserva Natural Otamendi, which was only 8km east of our hotel. This massive wetland was created in 1990, covering 2,700ha and is situated on the banks of the Parana De Las Palmas River. Long-winged Harriers simply thrive in these low wetland areas close to marshes, reedbeds, lagoons and wet ditches, so my hopes were high as I entered the Reserve. For such a large raptor, Long-winged Harriers surprisingly forage on a wide range of prey sizes and weights, ranging from insects at only 1g to a high of 450g in the case of White-faced Ibis. They also take small passerines together with rodents and young hares when hunting over cultivated fields and pastures.

As it was 8.30am I had not expected to wait until 9.10am to encounter my first

A majestic adult female Cinereous Harrier   *Josh Larsen*

Long-winged Harrier, which was an extremely large and lanky looking female with noticeably long wings, slowly quartering the marsh. This sighting was then followed by my first male which appeared similar in size and plumage to the female, but when he passed quite close to my balcony observation point I could clearly see that his upperparts were black, whereas the female's were dark brown. There is apparently much variation in size with this species, but it is still regarded as the largest of all the harriers, though by no means the heaviest. My first impressions were that these were hugely impressive birds with extremely long wings, which seemed to fill your binoculars when you engaged one to view, even at long range. On the other hand, when they came reasonably close you really did not need your binoculars, due to their sheer size and slow movement over the ground. At this stage it was hard to tell if the birds were breeding, though we did observe several brown and heavily streaked juveniles in the air together.

We remained in the Reserve until well after noon before retiring for lunch, with the main topic of discussion being Long-winged Harriers! On our return, we viewed the Reserve from an unsealed road close to the Parana River, which gave me close and personal views as the harriers passed directly overhead when they visited another large reedbed which lay directly behind us. From here we could see males and females carrying food to what were quite clearly nests amongst the tall reeds. On one occasion a male was seen carrying a large frog, with a female dispatching a large coot-like bird to another nest. I would love to have visited one of these nests, for obvious reasons, but they were well out in the marsh in treacherous habitat. Anyway, it was forbidden by the Reserve to enter the marsh at all times and I could understand why; as you had only to walk a few metres to become consumed in waist high water. Before leaving that evening, I encountered my first dark-phase birds, which could have been either males or females, but were clearly definable by their mainly sooty-black-brown plumage above and below.

The following morning 13th, while travelling to Otamendi, two males flew low across the main road in front of our car. One was returning to the Reserve with prey, the other setting off to hunt. We briefly followed the latter, but lost sight of him shortly after he entered a large field containing wheat. This was the first time I had observed this species away from their mainly wetland surroundings, which they seem to prefer more than dry areas. We once again returned to the unsealed road area where Chimango Caracaras often perched on bushes close to the road and occasionally pestered the harriers, as they returned to their main breeding grounds. Unlike the Cinereous Harriers, who were constantly bombarded by these scavengers, the Long-winged Harriers totally dismissed them when they appeared, their sheer size being an obvious advantage.

Our observations of the harriers in this area were more frequent than those from within the Reserve and I was able to photograph birds as they passed and was clearly able to identify whether they were males, females, juveniles or dark-phase individuals. The females rarely left the confines of the marsh. As we had witnessed earlier, all the hunting within the main part of the Reserve was clearly carried out by females, with males foraging over a wider area to procure food. Activity from within the marsh was in full flow that day, as food deliveries by males as well as females came in thick and fast to at least four nests. Prey was identified on two of the four occasions, with a rat and a small unidentified bird clearly seen, as two males passed my roadside vantage point on the way to their respective nests. What I did not like about this road was that when some cars passed by on their way to the Parana River, they drove deliberately at speed to create a dust storm, while others respected your presence by driving slowly. But this was the place to be, so we just had to grin and bear it when it occasionally occurred!

Cinereous Harriers food-passing   *Natalie Collm*

We returned to the Reserve for our afternoon visit and my observations were carried out from three balcony vantage points that I had utilised the previous day. During one brief sighting I watched a pair make a food-pass. The female swiftly returned to the nest which probably meant that she was either on eggs or had very young chicks. After two days of observations this was the first aerial food-pass I had seen. It was by no means spectacular, and could best be described as awkward; probably due to the large size of these birds. Overall, there were probably 15-20 breeding pairs at Otamendi, plus several unattached males and a good number of recently fledged juveniles, which were capable of hunting unaided. Certainly this was a good site that Sergio had chosen for me, especially as I had arrived at a time when nesting and fledging activity was at its peak.

An early start on 14th saw us leaving Campana, driving northwest towards Zarate. From there we drove north and firstly crossed the bridge over the Parana River, with a

Adult female Long-winged Harrier

Josh Larsen

second more highly elevated bridge to be negotiated over the Parana Guazu River, which officially marked the border between the provinces of Buenos Aires and Entre Rios. Once in Entre Rios province, we continued driving northward in what was now the Pampas, in search of more Long-winged Harriers. The large and lush green fields which were partially flooded in several places seemed never-ending and were usually filled with large numbers of cattle and horses, with Greater Rheas standing proudly amongst them. This was Gaucho country. These 'South American Cowboys', mounted on lean brown horses with expensive looking leather saddles and accompanying sheepskin blankets, could be seen tending their livestock. These cowboys were the 'John Wayne' replicas of Argentina!

When we reached a little village named Ceibas, we drove inland over dusty roads where Pampas Grass stood almost 3m tall in places. Here we searched the surrounding country-side for Long-winged Harriers, which had so far remained elusive. Twenty minutes or so elapsed before I encountered my first male and female. They were by no means numerous in this area and I realised I had been spoiled by the continuous activity at Otamedi. We also realised that very little foraging would take place over these grassy and wet open fields, since it normally takes place over areas where marshes and reedbeds exist. Several more harriers were encountered in this area before we retired for a well-earned lunch at Perdices, which is not far from the Uruguay River and the main border crossing into Uruguay itself.

We joined the Highway again (Route 14), and continued driving slowly southward, taking in different locations where Sergio had encountered harriers on previous visits. Again we were successful in observing several harriers, but this was a vast area and these birds could travel for miles before an observer would eventually engage one. Over the day, I observed more Roadside Hawks and Crested Caracaras than harriers, but I am not complaining, as these species and the Pampas, will never be seen in my native Northern Ireland.

Over dinner that evening we decided to visit the same area the next day, with one or two deviations expected along the way. We followed a similar route as before, but new territory was explored which proved fruitful. Other raptor species were also more noticeable that day and I saw several Savannah and Red-backed Hawks which were new species for me, as well as a tiny male American Kestrel. By 1.30pm we had eventually reached Perdices, where lunch and large beers (Quilmes) were had at the same roadside restaurant we had visited the previous day. But unaware to us both, our best harrier sightings were to come during our long drive back to Campana!

We rejoined the Highway, travelling south. It was actually a dual carriageway separated in the middle by a wide and sloping inward grassy ditch, with the occasional bush planted at various intervals along the way. To my total surprise, but not Sergio's, a male Long-winged Harrier was hunting over this ditch between both carriageways, which were busy at the time; with passing traffic going north to Uruguay or back south to Buenos Aires. Sergio had regularly travelled on this road in the past and had observed Long-winged Harriers foraging here on several occasions. Something he had mentioned briefly the previous day.

Despite the traffic, we managed to pull onto the verge to admire this male flying slowly and methodically northward and no more than 3m above the ground. We turned round at a nearby junction and began driving north again and minutes later I was able to photograph the bird from our moving car as it simply ignored me and carried on hunting. The bird travelled for at least 4km before it veered suddenly over a high hedge and into an adjacent field. This was an incredible experience and something you only dream about.

Dark-morph Long-winged Harrier   *Arthur Grosset*

Further north we managed to turn the car around once again and slowly drove south. When we arrived at the location where we first saw this bird, we pulled off the road to observe several birds of prey perching in a tall bush and a nearby telegraph pole, which we had not seen earlier. They turned out to be a Snail Kite and a Roadside Hawk. Another male Long-winged Harrier was also present. The harrier immediately took to the air and proceeded to hunt low over the ditch as its predecessor had done minutes earlier, once again travelling in a zigzagging northerly direction. On this occasion we did not follow but it continued on this course for some distance when it eventually went out of sight. Before leaving the area we crossed the busy carriageway to examine the ditch that these two harriers had been hunting over and discovered the ground was very wet and boggy in places where frogs were present. The previous day I had stated that Long-winged Harriers do not normally hunt in grassy areas. How wrong you can be! They are obviously great opportunists and these birds were well aware that prey was readily available here.

The 15th was my last day in Buenos Aires Province and so the morning was spent watching the harriers from our usual location just past Otamendi railway station. A strong wind was blowing which made the dirt road a very uncomfortable place to stand when passing vehicles failed to slow down. Harrier activity is usually best observed early in the morning as growing chicks are fed on a regular basis by both adults and today was no exception. I even managed to get several photographs of passing females as they flew to their respective nests, something I had failed to do previously. By 1.00pm the wind had strengthened and rain was in the air, which was preceded by thunder and lightning. By 1.30pm we were forced to abandon the area completely, due to torrential rain which lasted until after 4.00pm and so our final visit to Otamendi ended somewhat prematurely.

What a contrast the habitat was in the Pampas, compared with the parched ground and poor plant life in Patagonia. The former was obviously rich in prey and could easily sustain large numbers of breeding Long-winged Harriers, whereas the latter was lacking in both sufficient prey and good ground cover and these two problems alone appeared to regulate the small number of breeding Cinereous Harriers, which were present there. This was not just a trip to observe breeding harriers; it was a learning process as well!

David Hollands informed me that he and his wife Margaret were flying out from Australia in February 2007 to visit family in Buenos Aires. Since three years or more had swiftly elapsed since I last visited Argentina, it immediately rekindled memories of my successful harrier trip in 2004. As their holiday was for three full weeks I expected they would travel extensively, and I hoped they would observe Cinereous and Long-winged Harriers.

On 7th March 2007, I received a most welcome email from David on his return. It stated that he had encountered several Cinereous Harriers in the Torres del Paine National Park, in southern Chile. One particular pair caught his attention and during a discussion with the local bird guide (Paula), she told him that they had their nest in a native Lenga Beech tree – and not on the ground as is normal for this species! This was apparently in response to severe ground pressure from large numbers of Grey Foxes. Although she gave David vague instructions as to where the nest was, he could not find it. He suspected that the young had already fledged and this is probably correct, for when I was in Patagonia in early December 2004, there were already 7-10 day-old chicks in one of the nests I found.

Being unable to access a fax machine, I eventually found an email address for the Las Torres Hosteria where this guide is employed. Unfortunately I never received a reply to my urgent questions. This potential record of tree-nesting Cinereous Harriers, therefore, sadly remains unsubstantiated. Both David and I remain optimistic and although he cannot prove if the guide was right or wrong, her information sounded completely plausible to him.

Adult male Northern Harrier

George Jameson

10

NORTHERN HARRIER
Circus hudsonius

The **Northern Harrier** formerly known as the Marsh Hawk is North America's sole representative of a genus, *Circus*, whose 16 species claim territory on every continent except Antarctica. They breed from northern Alaska and Canada, south to Baja California, Mexico and the southern US, with the exception of several southeastern States. Nesting generally occurs in marshes, grasslands, meadows and cultivated fields and occasionally coastal sites are occupied. Winter movements occur from southern Canada, to almost all the southerly states, with birds also known to frequent the northern parts of South America. Communal roosts on the ground are quite common in agricultural fields, grasslands and saltmarshes.

On 3rd January 2006, I left a cold and wet Belfast for a nine-day birding trip to the sunshine state (even in winter) of Florida, where I would be visiting several of the state's prime wildlife refuges, in search of my first Northern Harriers. Amazingly, this would be my first ever visit to North America, even though the state of Florida is a popular holiday destination for Northern Ireland families. I would be based for part of my trip at my host's (Josh Larsen) home in Delray Beach, which is just over an hour's drive (north), from Miami International Airport and an ideal location for visiting at least four wetland areas, including the world famous Everglades National Park. The latter was badly damaged by Hurricane Wilma (America's strongest ever hurricane), on 24th October 2005, and only certain areas of the park had partially opened prior to my arrival, so to gain full and unrestricted access for the purpose of studying Northern Harriers, was highly unlikely.

The main purpose of this hastily arranged trip was to monitor the daily movements, hunting behaviour and plumage details of wintering Northern Harriers and if possible communal roosting. On 4th January 2006, at exactly 10.45am, my first ever Northern Harrier, a female, was interestingly observed at Loxahatchee NWR, in southeast Florida. The following day at the same location, a female with totally different plumage was observed hunting at 4.50pm, and in the process, she deliberately flushed a small flock of White Ibis, when she suddenly appeared over a tall stand of Sawgrass. It was clearly noticed that when several Black and Turkey Vultures flew low over the same area minutes later, the White Ibis flock and other avian species remained firmly on the ground. Only when the harrier was present did full-scale panic occur.

My only other visit to Loxahatchee was a four-hour afternoon visit on 9th January and this time the area was more fruitful, with a total of three seen, which were possibly the same two females observed previously and an elusive grey male. One of the females made a half-hearted attempt at catching a Boat-tailed Grackle, and then moved on to harry a small flock of Red-winged Blackbirds. As it was late afternoon (4.30pm), all three harriers had probably eaten well, before going to ground somewhere within the Refuge. Clearly this was a raptor solely intent on wreaking havoc, irrespective of the size of its quarry due, to its seemingly fearless nature.

Loxahatchee was also badly damaged in October 2005, by Hurricane Wilma and this was reflected by the amount of fallen trees and the low numbers of wildfowl and other relevant bird species, in comparison to previous winters. With very little prey available, this was probably why so few harriers were present at this usually prime location. At two much smaller wetland centres, Wakodahatchee and nearby Green Cay, bird numbers were also low, but the best and more natural habitat for observing Northern Harriers, was undoubtedly at the much larger Loxahatchee. Sadly, I observed no harriers at the former two locations, although a male had been seen sporadically at Green Cay.

One thing was now certain, if I wanted to see larger numbers of Northern Harriers, my host and most able guide Josh Larsen, suggested that we should move further north to Merritt Island NWR, in central east Florida. Prior to my arrival in Florida, Josh had already contacted several of the staff at Merritt Island regarding Northern Harriers, and was politely told that there were good numbers present on a daily basis. So, this was the place I needed to be! To my utter amazement, I immediately discovered that Merritt Island shares a large boundary with NASA's Kennedy Space Centre and Cape Canaveral, where the space shuttle and the historic moon launches took place.

At 8.30am after an almost three-hour drive north, from Delray Beach, we arrived at Merritt Island, for a four-day visit. We eagerly entered the Black Point wildlife drive, which is a seven mile, one-way, self-guided tour, through salt and freshwater pools and marshes. After only a short distance I soon discovered that Northern Harriers and other birds, were seemingly abundant here, with around 16 harriers observed by lunchtime. The diverse habitat was superb for foraging and roosting and on two separate occasions I witnessed two females boldly ambushing a small raft of Blue and Green-winged Teal, that were innocently feeding in the nearby pools. Although no prey was caught on either occasion, these harriers hunted in a fashion that was more akin to Western Marsh Harriers. Perhaps, its former name (Marsh Hawk), was more than appropriate for this harrier species.

Other harriers were noted hunting over and along both sides of a line of Mangrove bushes and also in open grassy areas, in pursuit of small mammals and birds, mainly Red-winged Blackbirds, which appeared to be plentiful throughout the Refuge. This regular quartering in areas where Mangrove was abundant was superb to watch, as they weaved effortlessly from side to side about 1-2m above the hedge-like bushes and when prey was detected they would instantly drop to the ground (with legs fully extended), and capture a rodent within seconds.

A pre-arranged meeting that afternoon, with Dorn Whitmore (head of public use), and Marc Epstein (the biologist supervisor), soon lead me to acquiring a permit for my harrier studies within the Refuge, with access also granted before sunrise and after sunset to give me a good chance of locating roosting harriers. Unbelievably, roosting Northern Harriers, had never been recorded here by the staff or the general public, but I firmly believed the habitat was ideal for a roost site, given the vast amount of prey available and the adjacent saltmarshes, which provided vital cover amongst the tall stands of cordgrass (the main constituent), with lesser amounts of Knot and Salt Grass also available.

Just before 5.00pm, when passing stop # 9, I immediately noticed several harriers

Top: Adult male Northern Harrier   Bottom: Female Northern Harrier

W. S. Clark

milling over an area of grassland to the east of the main drive and only a short distance from the Cruickshank Trail. The distance from my observation point to the nearest harriers, was approximately 500m and in good light (with the sun directly behind me), the birds could be clearly seen arriving at the site. The harriers were occupying a huge saltmarsh, which was completely covered in 1 metre high cordgrass. Several birds had already pre-roosted on posts and mounds of grass and at one stage five and then eight were seen circling in unison above the site before suddenly plummeting to the ground.

By 5.58pm at least 31 harriers were counted utilising this particular piece of ground, with my observations eventually ceasing at 6.30pm. The roost consisted mainly of adult females and juveniles with very rufous underparts, and two grey males. A few paler (and much older juveniles), with creamy unstreaked underparts were also recorded. The following morning (7th), my next watch began at 6.30am, with the first bird leaving the marsh at 6.56am, and the last at 7.06am. When I ceased watching at 7.30am, 27 harriers had left the area in various directions, with several passing directly overhead on their way to the nearby wetlands, to commence their daily hunting forays. By mid-morning, I had reported this unique find to Nancy Corona (wildlife ranger), who kindly informed Marc Epstein later that day.

During the evening count, which commenced for me at 4.30pm, Marc kindly joined Josh and I at the site, to see this previously unrecorded spectacle at Merritt Island. Our first harrier arrived at 4.37pm and the last at 6.01pm and similar to the previous evening several birds obligingly pre-roosted on the posts and mounds of tufted grass, before plunging to the ground in an abrupt 180 degree turn. Like the previous evening and the early morning watch, the harriers were heard calling excitedly as they arrived and left the site, with similar interaction recorded at other roosts, I have watched in the past. When we left the area at 6.30pm, a total of 25 harriers had settled at the site.

On Sunday morning (8th), my observations again commenced at 6.30am with the first bird hastily leaving the area at 6.59am and the last at 7.10am, with a reduced total of 20 seen, during a spectacular and frosty sunrise. It is not uncommon for harrier numbers to fluctuate between evening and morning counts, due to their irregular daily movements during the winter months, and with birds being disturbed by predators, this inevitably forces several, and occasionally, all of them to relocate elsewhere. A very rare Bobcat, was seen (the previous evening), with another heard nearby, so they may have been responsible for the reduced number. My final count that evening produced 29 roosting harriers, with the first arrival at 4.45pm and sadly my last at 6.02pm. On this occasion Jim Lyon (the bio-technician), joined Josh and I at the site for my final watch at Merritt Island. The striking grey male I had observed earlier in the day was not seen roosting here and over the three days, only four or five frequented the Black Point area, with several much younger males noted by their smaller size and eclipse plumage. Surprisingly, no males were recorded in the Peacock's Pocket area (several kilometres away), where up to 10 females and juveniles were recorded daily. Individuals at this location were observed hunting American Coot and wildfowl (small duck species) on the afternoon of 7th, and it seems reasonable to suggest that these harriers, occupied similar habitat each evening and may not have joined the main roost. As this Refuge, covers up to 140,000 acres, it is likely that other roosts exist elsewhere, due to the outstanding nature of the habitat and the abundance of prey.

After my three day sojourn at Merritt Island, I would estimate a minimum of 40 and a maximum of 50 Northern Harriers present at any given time throughout the day, with perhaps more roosting over the course of the winter months. Probably the largest recorded roost of Northern Harriers was at Fort Sill, Oklahoma in February 1985, when 303 were observed going to ground in a protected natural grassland covering 800ha. But the largest ever concentration at this site was on 7th February 1987, when 1,053 was tallied by 12 observers, which brings the numbers at Merritt Island into perspective. The vast grassland areas within the Refuge are control-burned on a five-year rotational basis to prevent alien species taking hold. The area containing the roost was last cleared during the winter of 2004/05. As a result, I have recommended to Marc and Jim, that when this area is due for burning in approximately 2009, that it is not done between October and March, as these tend to be the prime months for roosting harriers in the Western Palaearctic.

Finally, I must mention that the wintering Northern Harriers, I had the pleasure of studying on Merritt Island and elsewhere in Florida, was my 13th harrier species observed out of 16 recognised species and one subspecies, worldwide. My lasting impressions of them were that they were not dissimilar in size and body proportions to that of the Hen Harrier's I have studied so closely in Northern Ireland, since 1986. Until recently, the Hen and Northern Harrier, were classed by taxonomists as one species, but thankfully my harrier colleague Rob Simmons (South Africa) and Dr Michael Wink (Germany), split them – and they are now regarded as two separate species, thanks to the advances in DNA testing.

With regard to plumage details, adult males have dark slate-grey upperparts, with cinnamon spotting on their white underparts and underwings, whereas male Hen Harriers, are paler grey above and plain white below. Females appear slightly darker brown above than their European cousins, with juveniles having chocolate brown upperparts and very rufous unstreaked underparts. Juveniles sporting pale creamy underparts are a typical feature of older individuals, particularly as spring approaches. Due to their darker plumage, the buffy edging surrounding the facial disks of females and juveniles, plus the white uppertail coverts (or rump patch) are very distinctive, with the latter only visible when the harrier is seen quartering the ground for prey.

One abiding memory of my trip to Florida must be the unexpected arrival of over 30 Northern Harriers, to a previously unknown winter roost site, during my first evening at Merritt Island NWR. A close second, was watching the amazing skill and agility of these birds, when it came to hunting several species of duck in the nearby pools, it was totally awesome. Lastly, I observed my first ever Bald Eagles, the National emblem of North America, which was a fitting conclusion to a wonderful and definitely interesting visit.

Adult female Swamp Harrier, nest building

David Hollands

11

SWAMP HARRIER AND SPOTTED HARRIER

Circus approximans and Circus assimilis

The vast continent of Australia, which is mainly flat, has only two species of harrier, the Swamp and the Spotted.

The **Swamp Harrier** is regarded as one of the largest and also the heaviest of the world's 16 species of harrier, with some females weighing as much as 890-950g. It is found throughout mainland Australia, with breeding strongholds also in neighbouring Tasmania, New Zealand and to a lesser extent on several other offshore islands. As its name suggests, it is usually seen frequenting marshes, reedbeds, or open pasture near water and even over large fields containing a variety of crops such as wheat, barley and rice. Nesting generally occurs on the ground but there are several records of this species tree-nesting in New Zealand!

The **Spotted Harrier** is also widespread, but sparsely distributed and on occasions highly nomadic, especially in drought years when prey is extremely scarce. There are also small breeding populations in the Lesser Sunda Islands, Celebes, East Timor and Sulawesi. In contrast to the Swamp Harrier, it tends to favour the dry inland grassy plains further north, where cereal fields and open wooded areas exist. It is occasionally seen hunting between the trees and habitually nests in woodland edge trees, and totally avoids dense forests. This species is the original tree-nesting harrier!

Almost four months had elapsed since I met Drs David and Margaret Hollands, for the first time on 15th June 2006, during their all too brief visit to Northern Ireland. David and I had been regularly corresponding about both species of harrier since 1994 and during our 'get-together' it was decided there and then that I should visit them in Australia in early October, for the purposes of studying both species during the breeding season. I was absolutely thrilled by the prospect of travelling this huge distance (11,500 miles), to see two more of the world's harrier species!

On the morning of 5th October 2006 at 5.30am, I left a wet and misty Belfast for the long and tedious journey to Melbourne via London and Hong Kong. After a two-hour break in Hong Kong, I eventually arrived in Melbourne at 8.00pm on the 6th, after flights which lasted some 21 hours. Thankfully David and Margaret were there to meet me after my exhausting journey and luckily for me I would be spending my first night in Australia at their flat in Melbourne, where I was to appreciate a good night's sleep.

The following morning, I had another long journey ahead of me as we left (by car), for the Hollands' family home in Orbost, which is 376km southeast of Melbourne. Around half distance we drove into a picnic site a couple of kilometres east of Rosedale, for a welcome coffee break and at precisely 11.15am I had my first sighting of a female Swamp Harrier. The bird had apparently risen from a partially flooded field adjacent to the picnic site and although the sighting was brief it was a foretaste of things to come over the next 12 or so days. Then, at 5.05pm within five kilometres of David's home, we briefly stopped at a small marsh which was densely populated with tall Phragmytes and Willow bushes. Within seconds of alighting from the car my first male Swampie took to the air and

Spotted Harrier – the original tree-nesting harrier

David Webb

quickly disappeared behind a tall stand of deciduous trees. A pair of Swamp Harriers was in the early stages of nesting but on this occasion there was no sign of the female. This particular site would be highly significant for historic reasons, to both David and I during the course of my studies!

The next morning (8th), we left early for an area known as Marlo, which is situated between the famous Snowy and Brodribb rivers. Unlike the previous day, when it was very warm and sunny, it was cool and extremely windy with a few drops of much-needed rain in the air. We stationed ourselves by the roadside. A small lake surrounded by reeds was to our left, and a flooded area with much shorter reeds to our right. Open green fields containing cattle were in the background.

Within minutes of our arrival we savoured quite a few male Swamp Harriers, as they drifted effortlessly into a strong headwind in search of mates. Then the females began appearing with the males vying with each other for instant selection. The views we had of these birds were tremendous as they continually crossed the main road, with males hotly in pursuit of females. On several occasions it became clear that quite a few of these birds had already paired off, with males suddenly landing amongst the reeds quickly followed by the respective females. These were very large harriers, particularly the females with their long wings and heavily built bodies. The males were noticeably lighter and slightly smaller. Over two hours of observations revealed that at least six pairs were present, all in the mid stages of courtship prior to nesting, although no sky-dancing, was observed on this occasion.

We again visited the Marlo area the following morning with the weather in total contrast to the previous day; it was now bright, warm and sunny. David had brought his camera along hoping to get good shots of the birds as they passed by, but sightings were irregular and it was also difficult to get reasonably close to the birds. On one occasion, a male came by with a female in close attendance and when another male appeared on the scene, the first male attacked the intruder, with both birds briefly talon grasping and tumbling for a few seconds, before the second male retreated from the scene. In another part of the marsh a female perched for long periods on a post close to the reedbed and when a male appeared, probably her mate, she persistently flew to a grassy area with the male landing beside her. She immediately offered herself to the male on at least two occasions, lying flat on her belly with wings outspread, but copulation did not occur on either occasion. Before leaving we did see a single male carrying nesting material to a potential nest site with the female also landing in the same area, but both birds were in the air again within seconds and soon drifted away from our observation point. It was quite clear that nesting had not commenced yet, but was probably only days away.

The afternoon was spent at Newmerella Hill. This was the area we had briefly visited on the 7th, which was only a short distance from my host's home. Earlier that morning we had received a phone call from a friend of David's (Tony Mitchell), who had seen the female and male carrying sticks into the swamp. A hide was quickly erected on a hillside which overlooked the area, and from 1.30pm until 4.00pm I diligently watched the site.

Adult male Swamp Harrier

W. S. Clark

The historic significance of this particular site dated back to around 1980, for it was here that David photographed breeding Swamp Harriers for his superb book, *Eagles Hawks and Falcons of Australia*, which was published in 1984. Now, a quarter of a century later, I was getting the opportunity to observe these birds from a discreet distance. During my two and a half hour stint in the hide, I only made eight observations, one of which saw the female arrive with a short stick firmly clasped in her talons at 1.58pm, with the remaining sightings all featuring the male, soaring and checking out what was almost certainly a nest site.

The reason the hide was placed well away from the nest site was explained to me prior to my arrival at Newmerella. The Swamp Harrier is regarded as Australia's most sensitive raptor, as nesting females have been regularly known to desert their eggs and young even after the briefest of human intrusions. The males are also easily disturbed and I did notice during my first encounter at this site and on arrival that afternoon, that he was off instantly at the sound of a car stopping or the door opening and closing. On occasions, they have accepted a hide at the nest but they are extraordinarily shy and wary, whereas the Spotted Harrier is much more approachable and readily accepts a hide. I was unaware of this situation and I know of no other harrier species that reacts in this manner.

The mornings of the 10th and 11th, were again spent in warm sunshine at the Marlo area. Once again from David's perspective, photography was extremely difficult, but how I enjoyed watching these large harriers with almost similar plumage details. Both had rufous and streaked underparts but the mainly brown upperparts, especially on females, were replaced by extended grey patches on the male's upperwings close to the primaries. In poor light it was difficult to separate males from females, particularly in certain flight sequences. All our photography and observations of these Swamp Harriers at Marlo, were carried out from within the spacious confines of our 4x4 jeep, for fear of causing undue disturbance to these birds during their pre-breeding displays.

During both days, several, but not spectacular aerial food-passes were noted, and at least one pair had commenced their sky-dancing. The birds kept frequenting a small fenced-off area of reeds close to the roadside. The male would begin his undulating routine by climbing high into the air with, on one occasion, the female nearby. He would land amongst the reeds with the female following suit and then suddenly both were in the air again. This behaviour was repeated on at least three separate occasions. It is likely that copulation occurred briefly when both were hidden amongst the reeds, although this was not actually observed by either of us. In one sequence during his frenzied display, the male completed at least two 360 degree rolls, before swinging up out of his dive. This I have never witnessed before even though I have spent many hours observing Hen Harriers sky-dancing! In males, these aerobatic displays highlighted the contrast between their pale rufous underparts and dark brown upperparts, with each being flashed continuously during the frenzied tumbling routine.

It is also important to mention, that on only two separate occasions did I observe the resident Swamp Harriers at Marlo away from their their wetland quarters. This occurred

Newmerella, where I watched Swamp Harrier at the nest   *David Hollands*

on the afternoon of 10th, when a male was observed hedge-hopping, close to the main road and on the morning of 11th, when a female was mobbed by two Australian Magpies, whilst quartering a large open field. Although it would have been superb to see this species tree-nesting, there are no known records for Australia. In nearby New Zealand, however, nests range in height there from 2.5 to15m, in a wide variety of trees and shrubs, but definitely not in Sitka Spruce! Other notable highlights over these two days included, at least two Whistling Kites and a massive adult Wedge-tailed Eagle, which were both new species for me!

The afternoon of the 10th also brought me in contact for the first time with one of Australia's most deadly snakes. Australia has an abundant and a wide variety of wildlife, but it also hosts several of the world's most venomous snakes and creepy crawlies, such as the infamous Funnel-web Spider. While out walking in a wetland area with David and Margaret, Margaret's foot was within inches of treading on a basking Mainland Tiger Snake, reckoned by many, including David, as Australia's and one of the world's 10 deadliest snakes. A very slow and cautious retreat from the area was carried out immediately and everyone escaped unscathed, as the snake thankfully refused to uncoil from its slumber. That was my first and last encounter with such a beastie during my holiday, but that moment made me aware of the hidden dangers that await the unwary in out-of-the-way places.

On the 12th, we all left Orbost, to travel back to Melbourne as Margaret was leaving the following day for Manchester, to visit her sister who had been taken ill prior to my arrival in Australia. This journey takes at least five hours, but that day the heat was overpowering with Melbourne experiencing its hottest October day in almost a century with the temperature peaking at 36.6 degrees at 4.33pm. In fact I had arrived in the State of Victoria, during the worst drought that southeast Australia had experienced for decades with strong northerly winds fanning more than 250 fires across the state. The following afternoon we would travel up into New South Wales, to monitor Spotted Harriers. This area was apparently suffering greater hardship than neighbouring Victoria.

At around 2.45pm on the 13th, we left Melbourne International Airport, for the long drive up to our destination, Coinnee Woods, north of the town of Jerilderie, in New South Wales. As we passed over the Murray River which divides both states, it soon became evident how badly the lack of rain this spring had affected this part of NSW. This area is commonly known as the Riverina, which is rich in many varieties of fruit and cereal crops that grow on the vast open plains and during the spring they are usually well refreshed by the rains and the vast irrigation system that exists there. Sadly, this was not the case in 2006, with farmers allowing sheep and cattle to graze amongst crops that simply did not reach fruition. The drought had also affected the numbers of Spotted Harriers breeding here, with none observed on the way and only one known nest for me to observe so far! In good years, as many as 15-20 pairs breed from Mildura in northwest Victoria to Wagga Wagga (pronounced Wogga Wogga), in eastern NSW, even though this species is inclined to live a nomadic lifestyle and rarely returns to the same area to breed each season.

Female Swamp Harrier in its familiar wetland habitat

Arthur Grosset

The next morning we were up bright and early and on the road by 6.45am to meet up with David Webb, who had found this precious nest. David who lives at Griffith, was also a considerable distance from this nest site, which was slightly east of the non-existent Lake Urana. He met us at a nearby road junction and from there we proceeded to the site. It was here that I viewed my first ever tree-nesting Spotted Harriers!

At 8.00am, I had my first sighting of a Spotted Harrier. The male conveniently flew from the nest tree, where he had probably been delivering food to, what I had been told, were two large chicks. Minutes later, the much larger female appeared. Both birds gave us good views as they continued to encircle the vicinity of the nest site. 'What a pair of beautiful birds', I immediately remarked to the both Davids, just after the birds had disappeared from view! These birds made the hair stand up on the back of my neck and until then I had regarded the endemic Black Harrier from South Africa, as my favourite and the most attractive of all the 16 species worldwide. After viewing both these birds at close quarters, I could immediately understand why the Spotted Harrier, is regarded as Australia's most beautiful diurnal raptor, even though there are another 23 contenders in line for that prestigious title! So rare are these birds to southeast Victoria, where David Hollands lives, he has only encountered two, in the past 43 years!

As the birds had completely disappeared I made my way to the nest site which was approximately 100m in from the main road. The tree was identified as a White Cypress-Pine *callitris sp.*, which was probably 15m high, with the nest situated on the extreme but strong outer branches about 6.5m above ground. The ground directly below was heavily soiled with white crystallized excreta as was the sides of the nest and the surrounding branches. A few fresh pellets also lay on the ground as did the remains of a single hatched eggshell, but surprisingly no prey items and no moulted feathers. Looking up and also photographing the nest for record purposes, I could see that it was sturdily built and in no danger of collapsing unlike some of the Hen Harrier tree nests I have encountered in Northern Ireland, in recent years. The nest appeared oval in shape and was at least 60cm long and 30cm high, although from below it looked not much larger than a normal-sized dinner plate. The nest material of mainly twigs, grass and eucalyptus leaves were clearly visible with the latter appearing to be fresh and probably replenished each morning, by either the male or the female.

Interestingly, the Spotted Harrier is considered to be a primitive member of the genus *Circus*, because of its unusual plumage details which bears a considerable resemblance to that of the Serpent-eagle *Spilornis spp* and its unique habit of nesting in trees. Since their nests are flimsy and unusually flat structures compared with ground nests of other harrier species, and because the young prefer to roost on the ground after fledging, the Spotted Harrier has only recently (in evolutionary terms), become arboreal and is therefore derived from the typical harriers, rather than ancestral to them.

Minutes later as I retreated from the site, the male appeared with one leg partially dangling as if he was carrying prey and when I mentioned this observation to DH, he told me that it was a characteristic feature of this species. I had actually observed the

Female Spotted Harrier feeding her chick at a tree nest

David Hollands

female doing this earlier in the morning and immediately thought she was carrying prey but I passed no remarks to DH or DW, at that particular time. Occasionally, both legs are slightly lowered in flight, but I did not record this during my intensive observations of this species.

Similar to the circumstances at Newmerella, when I observed Swamp Harriers, a hide was hastily erected a discreet distance from the nest site and I spent the next three and a half hours watching these birds and their chicks. From 11.05am to 2.35pm, I made 29 separate observations at this site which included the discovery of three chicks (and not two as first thought) at 1.02pm, with prey deliveries by both male and female at 1.27pm, 1.35pm and at 2.00pm by the female, when she fed the youngest of the brood. Prior to the three prey deliveries, the three chicks had been heard uttering faint begging calls and obviously both adults had responded immediately. Activity on the nest by the eldest two chicks in particular, was virtually non-stop, as they stretched and wing-flapped. All three chicks were fully-feathered, with slight traces of down on the head of the eldest two, whereas the youngest chick's head was fully covered in down and also on the breast and belly. The two oldest continually preened, with the down floating off in the wind or clinging to the edges of the nest as the baby of the brood looked on in a somewhat confused manner.

Only when you were really close to the adults (as I was), could you appreciate not only their size, as these birds were on a par with the Swamp Harrier, but their much lighter bodyweight, and slight but clearly definable variation in plumage colouration between male and female. The male is much darker above and below, but the silvery grey upperwings and chestnut-spotted white underparts with a prominent black and white barred tail, on both species, are unmistakeable in the field. The chicks were also attractive subjects in their own right, with their rufous-brown upperparts and buff underparts also observed! With regard to the chicks, they did show signs of general unsteadiness while standing upright on the nest, but this species has a long history of tree-nesting, unlike Hen Harriers, and at no time did they appear in danger of falling to the ground. This had been a long and exciting day in the field and so we gracefully retired from the area around 4.30pm and slowly made our way back to the home of our hosts, Gwen and Murray, in the expectation that we might locate another pair of Spotted Harriers. Sadly, for me, none were observed!

Sunday the 15th was another beautiful and warm morning, as we left Coinnee Woods at 7.15am to drive the 83km to the nest site. On arrival, there was no sign of the adults and there was no activity on the nest. Hopefully, the two larger chicks which had been very active the previous day, had not fallen during the night. Today, we decided to use our 4x4 vehicle as a hide rather than being conspicuous, and so this was parked at what David H regarded as a safe distance from the nest site. After 45 minutes of failing to observe either adult and still no movement on the nest, we decided that the best policy was to check the base of the tree for dislodged chicks. Thankfully, none were found and so we hastily retreated back to our vehicle. It was probably my pessimism that caused this slight panic

Spotted Harrier watching over us at Urana, New South Wales

David Webb

in the first place, as chicks of similar size in a Hen Harrier tree nest would surely have been on the ground after the constant activity I had observed the previous day. There again, Spotted Harriers are experts at arboreal nesting and I must stop comparing their unique traits with those of my local Hen Harriers!

Shortly after returning to our 4x4, the chicks began to stir on the nest and soon all three were accounted for. Our first prey delivery of the morning did not occur until 10.45am, when the female suddenly appeared from behind the tall pine carrying, what I thought at the time, to be a mouse. She flew in over the nest again at 10.52am with as before, one leg dangling slightly, while the other firmly held the prey item. At exactly 11.00am, she approached the nest for the third time and we could see a long tail dangling which belonged to the body of a good-sized rat. On landing, our line of vision was slightly obscured by a protruding branch and we could not see her feeding the chicks. When she retired to a nearby tree, close to the main road, she was mobbed by several noisy Galahs.

There were at least three trees here which both male and female utilised to observe their chicks. One was a favourite outpost for a large flock of Galahs. They took exception to the harriers perching in this tall tree by the roadside, especially when they decided to remain there for long periods. On several occasions, these cheeky cockatoos would mob the offending harrier from a safe distance by calling loudly as they passed over it in small numbers and also by perching as close as safely possible without being attacked. It was a complete waste of time on their behalf as the male and occasionally the female, totally ignored their advances. Another favourite tree, particularly used by the female, was a short distance from our vehicle and it was here we both photographed and observed her radiance on numerous occasions. It was quite busy around the nest site that day. Not only were large numbers of Galahs conspicuous, but a pair of Nankeen Kestrels hunted in the area for several hours as did an all too brief Australian Hobby.

There were two further prey deliveries in the afternoon which were small birds, about the size of a Richard's Pipit. Owing to the drought in this area, the harriers' most favoured bird prey during the breeding season, the Little Button-quail, was apparently absent from the area. After two full days of observing the site closely, no more than six items were delivered to these huge chicks, reflecting the shortage and diversity of food available during that season. From my observations at Hen Harrier ground and tree nest sites over the past 22 years, chicks of this size would be fed almost on the hour every hour by both adults until fledging occurred.

The previous day, the 14th, David Webb told me that he had observed a male Spotted Harrier in the Oaklands irrigation area, southeast of Jerilderie a few days earlier on the 11th. We therefore decided to travel in that direction on our way home. When driving along this straight and unsealed track, a large bird was observed at 5.00pm, coursing low over a wheat field to our left and when our vehicle came to a sudden and premature halt, we discovered that the bird was a large female Spotted Harrier! When she came quite close to our jeep she suddenly dropped down into the main channel that provides the water for the crops in this area and then suddenly took to the air again pursued by several raucous Australian Magpies. This immediately told us that there was probably a second pair breeding in this area, but as tomorrow was our final day here we would not have time to look for the nest site, which was presumably nearby. I was elated at this find and it was also good to know that the pair of Spotted Harriers I had been monitoring earlier was not the only pair in this region.

Spotted Harrier landing at its tree nest   *David Webb*

Our final half-day was again spent at the Urana nest site and when we arrived at 8.15am, the youngest chick of the brood was amazingly, fighting with one of its elder siblings. It was giving this much older chick a right 'going over', as it persistently bit the top of its head and neck on umpteen occasions, before retreating to the far corner of the nest when tranquillity was finally restored once again. Both adults were conspicuous in the area but neither had delivered food to the chicks by the time we left for Orbost at 12.30pm. David W arrived prior to our departure and I thanked him sincerely for sharing his discovery and for telling David H about this nest. Without his help I would undoubtedly have been chasing shadows around these vast plains in pursuit of very few pairs of Spotted Harriers. Obviously I would have preferred to monitor several nests at different stages of the breeding cycle, but serious drought in the area prevented this. I was thankful however to have had the opportunity to observe the Urana site. After over 20 hours of observations at this nest site and the surrounding countryside, I could now fully understand why Spotted Harriers nest in trees. Quite simply, because there is inadequate ground cover! A final twist was about to occur as we made the long journey back to Orbost, via the scenic mountain route!

As we were descending from Mt Hotham and a short distance from the town of Omeo late in the afternoon, David H's mobile phone unexpectedly rang! It was David Webb, who now relayed to us that he had found another Spotted Harrier's nest 30km east of our original site and not far from the town of Lockhart. This nest contained much

Spotted Harrier tree nest   *Graeme Chapman*

younger chicks with both adults ever present, but sadly we could not return to monitor the site as we were well over half-way to Orbost. What a pity this nest had not been discovered earlier that morning as we could have made a slight detour to include it prior to our departure. It would have been madness to return, but I must admit it was highly tempting!

Within ten days of my return to Belfast, David Webb informed me by email that two of the three chicks at the Urana nest had fledged safely by 26th October, with the third imminent. A follow-up visit on the 29th confirmed that all three had fledged, with one perching on the edge of the nest and the other two in neighbouring trees. In total, David Webb managed to locate four breeding pairs between the towns of Urana and Jerilderie, all of which produced young – two each from the three remaining nests and all in similar trees to that of the first nest discovered.

Spotted Harrier tree nest with chicks   *Graeme Chapman*

Sadly, the Newmerella Swamp Harriers were not successful and shortly after Christmas I was told by David Hollands that the nest had failed. David had been purposely monitoring this nest on a regular basis and when he had not observed either adult bird for a couple of weeks he decided to investigate further. Treading carefully through the swamp, the nest was eventually located and there he found the desiccated remains of three partly-feathered young. Only wings, some feet and a beak remained and many of the bones had been broken. A fox was suspected, but it is hard to imagine that a chance visit to a dense and dangerous swamp was the cause of this predation.

Initial confusion can occur in identifying some species of harrier, although there were really none that should have given me any bother. I did however, misidentify a Brown Goshawk as a Swamp Harrier, on two occasions, as it flew low over a grassy paddock, in 'harrier-like' fashion. When corresponding with raptor expert Bill Clark in the USA, I discovered that he had experienced a similar problem with Red Goshawk during his visit, (a species that I did not observe). He too was struck by how 'harrier-like' they were in flight and overall shape. It was reassuring to know that I was not the only person to suffer from identification problems when observing a species for the first time.

Before bringing this chapter to a close, I had hoped to meet up with two people prior to leaving Australia, Lindsay Cupper from Mildura, who with his late father Jack, had photographed Swamp and Spotted Harriers and featured them in their classic book, *Hawks in Focus* (1981), and also David Baker-Gabb from Melbourne, who had diligently studied the breeding ecology of both species for many years. Unfortunately, Lindsay was away filming breeding parrots, though I did manage to speak to him by phone. I was however successful in meeting up with David B-G for dinner on my last evening in Australia, when harriers were the main topic of our conversation – and why not!

The aim of this trip was to study both Swamp and Spotted Harriers, and this was successfully achieved, but this would not have been possible without the help of several absolutely brilliant people! I must admit, that there were other worthwhile distractions that David Hollands had lined up for me, such as watching his favourite species of raptor, Powerful, Sooty, Barn and Southern Boobook Owls.

If I was to take anything positive away from this trip it would probably have to be, that I actually rekindled David H's interest in going out to photograph Swamp and Spotted Harriers once again, something that he had not done for at least 25 years! That is the effect harriers have on mere mortals like David and I.

Adult male Papuan Harrier

Rob Simmons

12

PAPUAN HARRIER
Circus spilothorax

The **Papuan Harrier** is regarded as a resident endemic, with a fairly wide distribution from sea level up to 3,500m. It is known to frequent a variety of different habitats, from extensive areas of grassland, open floodplains, and aquatic vegetation fringing swamps and lagoons. Having a large hunting range it is usually rather scarce in the mid-mountain grasslands, though in areas where food is abundant it may be relatively numerous. Similar to other harrier species, prey is known to include lizards, small mammals, frogs and birds. The adult male is strikingly handsome with black, silvery-grey and white plumage, whereas females show brown upperparts and paler underparts with dark streaking on the breast; the tell-tale white uppertail coverts are visible on both sexes. Some females are known to retain the dark brown plumage of immature birds, which may cause considerable confusion in the field. In the highlands, a melanistic phase also occurs, with males described as completely black with a pale tail and prominent white uppertail coverts.

Papua New Guinea is the highest and second largest island in the world, exceeded only in size by the ice-covered terrain of Greenland. Geographically, it is bordered by Indonesia to the west, Australia to the south, the Solomon Islands to the east and Micronesia to the north. The island itself is split in two. Irian Jaya is to the west and New Guinea to the east, with the islands of the Bismarck Archipelago, the Admiralty Islands and Bougainville in the Solomon Islands, adding to this huge land mass. Nearly 85% of the island is still carpeted with tropical rainforest and is home to 38 out of 43 species of Birds of Paradise, for which New Guinea is world renowned.

For me personally, there were several reasons for undertaking this trip. Firstly, out of the world's 16 species and one subspecies, the Papuan Harrier was the only species I had not studied. Secondly, there is little known about the conservation and genetic status of this supposedly endemic raptor, and so it was probably best left until the final chapter. With no detailed studies of this species' breeding ecology since 1978, it also attracted the attention of harrier colleague, Rob Simmons with whom I had studied African Marsh and Black Harriers in 2002. Since 2006 we had been regularly corresponding and trying vigorously to get this expedition to Papua New Guinea not only organised, but sufficiently funded by sponsors to help with the considerable expenses involved in long haul and internal flights, together with accommodation in Papua New Guinea, for a two to three-week period. There were many highs and lows along the way, but thankfully the gloom lifted and all doubts dispelled, when the expedition was eventually scheduled to go ahead with the arrival of Rob in Port Moresby on 24th April 2007.

Another valid reason for this trip was to investigate the controversy that surrounds the very existence of this species on New Guinea. It has similar plumage to that of its nearest relative, the Eastern Marsh Harrier, which migrates to south-eastern Asia in winter and is believed by some, to frequent this unique island.

The status of another species, (going on current knowledge), the Australian Swamp Harrier is also uncertain as immature, sub-adults and females of the Papuan Harrier are also similarly plumaged and frequently misidentified as this species. But, the Swamp Harrier has been recorded with certainty in the southern swamplands of New Guinea on at least two occasions (1942 and 1964) and it is reckoned that some of the harriers which frequent the Waigani Swamp and sewage ponds near the capital, Port Moresby may well be this species (Coates1985). Unfortunately vital proof is lacking. Not surprisingly, there have been a few unconfirmed reports of the Papuan Harrier frequenting mainland Australia, but these sparse records remain unaccepted and are believed to be oddly plumaged Spotted or Swamp Harriers. A third species, the Pied Harrier, another migratory species from eastern Asia, has dubiously been recorded, but is highly unlikely to occur there.

There were five participants on this fact-finding expedition, namely myself, Rob Simmons from South Africa, his twin brother John (bird artist) from England and elder brother Michael from Java/Canada, with Leo Legra from the Wildlife Conservation Society in Papua New Guinea making up the final member of the team.

My long journey to PNG commenced on 28th April, with a short flight from Belfast to London Gatwick, where I stayed overnight with Rob's brother John who lives nearby in Kent. The next morning we left Gatwick for PNG, via scheduled Emirates flights to Dubai and Singapore, where we were met by Michael. From Singapore we flew with Air Niugini to Port Moresby, followed by an internal flight to our first destination, Mt Hagen, in the Western Highlands.

At 5.30am on 1st May we eventually arrived at Port Moresby International Airport, where a nearby tourist sign read – *Welcome to Paradise*! By 10.00am we had landed at Kagamuga Airport, in Mt Hagen, but not before I had savoured the awesome scenery from my convenient window seat. The mainly wooded hillsides rose up thousands of feet to meet the sky and as the plane banked and gradually got lower, villages were suddenly revealed, seemingly in the middle of nowhere, where crops, grassy areas with small lakes and rivers dotted the landscape. The whole area was so green it would not have been out of place in Northern Ireland, but it immediately reminded me of the habitats I had seen on Reunion Island, in the western Indian Ocean. But this was the South Pacific and most definitely by what I had observed so far, a beautiful and virtually unspoilt island – paradise really does exist for those who want to believe in it!

After 22 hours of continual flying, not counting the tedious hours spent waiting for our flights, all three of us were transferred to the Highlander Hotel, in Mt Hagen, where we duly relaxed, supplemented by cold beers and food while awaiting the arrival of our expedition leader Rob Simmons, our guide Leo Legra and driver Onika Okena. They were making the 115km drive up from Goroka in the Eastern Highlands to meet us, so that our Papuan Harrier activities could commence in earnest the following morning.

On 2nd May we left the hotel around 8.30am and hastily made our way through the busy streets of Mt Hagen city. I, in particular, was buoyant that morning as I had just watched Liverpool FC defeating Chelsea to reach another European Cup Final, so all I needed to complete a perfect day was some Papuan Harriers. Prior to his arrival in

A Papuan Harrier nest

Rob Simmons

Mt Hagen, Rob had been allowed access to the grasslands that encompassed Goroka Airport, by the manager there, as this is where he observed his first Papuan Harriers, an adult male, female and an immature female. Our priority now was to meet up with Rex Topiso, the regional manager of the airport at Mt Hagen, as he had also granted us unlimited access to the grassy and wet areas that were present either side of the 2.5km long runway. On arrival we immediately observed large numbers of Black Kites. We hoped harriers would be residing here too. This area was also an ideal location for an accomplished artist like John, as the high peaks of the surrounding and densely forested hills provided him with a scenic backdrop for his bird paintings, which I hoped would include the Papuan Harrier.

We were soon introduced to Rex and also Russel Palia, the electrical supervisor, who regularly patrols the runway to check that all the landing lights are fully functional. Then after a safety briefing we were escorted out onto the designated paths that surround the runway on all four sides. The airport here is 1,670m asl with aircraft of all sizes and helicopters regularly using the area on a daily basis. What a strange habitat for harriers to frequent! I was absolutely amazed that we were being allowed to access the inner sanctum of this airport in the first place, for if this had been in the UK or elsewhere in the world we would not have been granted permission due to the high security levels that exist at most of the world's major airports.

As we drove slowly towards the south end of the runway our attention was drawn to a bird perching on a box-covered light. Not surprisingly, our 4x4 drew suddenly to a halt and it soon became clear that the bird we were observing was an immature male Papuan Harrier. For the record, this my first Papuan Harrier, was observed at 9.30am, which meant that I had now seen all the world's 16 recognised species of harrier. I was heartily congratulated by my three colleagues, Rob, John and Michael! Rob, in particular freely admitted to me that this was only his ninth harrier species, which was no mean achievement by a person who was a highly eminent colleague, when it comes to his studies of the genus – *Circus*.

Over the course of the day we had consistent sightings of at least three to four different males, but only one was a fully-plumaged adult bird, the remainder being sub-adults, with probably only two adult females present and two to three juvenile or immature birds, one of which sported rufous thighs and underparts. During the course of my intensive harrier studies worldwide, I had never observed so many individuals in one day with such contrasting and variable plumage differences. Because of these plumage variations we gave these birds identification names for future reference – 'Two Spot', 'Smudge', 'Broken Primary', 'White Head' and 'Brown Back', to name but a few. Obviously our studies here were not easy – probably 'interesting' would be the correct term to use after our first day's observations.

At the time, I remarked that the plumage of the adult male was so reminiscent of a male Reunion Harrier, but that this bird was much larger with an exceptionally long tail. Remarkably, both these species breed south of the equator at 20 degrees below for the

Reunion Harrier, but only 5 degrees below for the Papuan Harrier, both being endemic to their respective islands. In fact, there are three other harrier species where the males sport similar plumages to those mentioned above. A third island endemic, namely, the male Madagascar Marsh Harrier, which also breeds between 12-25 degrees south of the equator, has similar plumage details, as have the male Pied and Eastern Marsh Harriers, which incidentally breed north and east of the equator. It would be useful to know if these five species are genetically linked, even though they vary in size morphologically. Perhaps one day in the near future, when all the scientific studies of the above named species have

A pair of Papuan Harriers   *Rob Simmons*

finally been addressed by Professor Michael Wink and Dr Robert Simmons, they will supply a definitive answer.

The following morning we were back at Kagamuga Airport around 8.30am and proceeded to drive slowly towards the north end of the runway. This area had not been fully watched for harriers the previous day and so I volunteered to carry out my observations there until noon. The remaining four observers and our driver concentrated on the middle section and south end of the runway where all the activity was noted 24 hours earlier. My first encounter with a harrier that morning was shortly after 8.35am when I noticed an adult female preening by the edge of the runway. At 8.50am, this light brown individual, with noticeable barring on the underwings, was spooked by the arrival of an incoming aircraft. When the plane taxied slowly towards its stand, the bird reappeared from the tall grass and resumed preening and wing-stretching in virtually the same place. On this occasion the white uppertail coverts were clearly observed.

When the bird took to the air and landed a good distance away in the long grass and out of my line of vision, I began walking back in the direction of the entrance. When I got to the top end of the path around 10.00am I suddenly noticed a very dark brown harrier in the air to my right. It was carrying prey but I had not observed the customary food-pass. Within seconds a lone sub-adult male appeared from behind a stand of tall bushes and I immediately identified it as 'Two Spot', as it clearly showed a large dark spot on each underwing. This very dark brown female suddenly landed amongst rank vegetation a short distance from me, and I assumed she had returned to her nest with the prey item.

Rob and Michael had also noticed these two birds and came to assist me in trying to locate a nest. Within minutes we had entered the long grass and as I got near to where I thought the bird had landed, she suddenly took to the air and flew northwards. On the ground lay feathers belonging to a small bird, which despite our best efforts, we were unable to identify. Sadly, there was no nest! A brief discussion about this very dark individual then ensued and we instantly recalled that we had observed both of these females the previous day around lunch-time in the same area. On that occasion the dark plumaged female attempted a brief display as both it and the paler female circled for five minutes directly above us. On reflection, this may have been no more than a territorial dispute rather than a pre-nuptial display.

The afternoon was spent around the centre and south end of the runway where the immature male with the very white head and breast that I had observed for the first time the previous morning, was now firmly ensconced on a four-legged metal stand approximately 50m from the path. Rob, walked to the stand and the bird flew off immediately. His efforts were rewarded with the retrieval of two fresh pellets. Other pellets were collected from the top of the box-covered lights and from below the two wind-socks at the north and south ends of the runway. Analysis of the pellets revealed that the harriers were catching small birds and mammals. The former were numerous in the area, namely – Hooded Mannikin, Pied Chat, Richard's Pipit, Tawny Grassbird, Brown Quail and Willie Wagtail. Later in the afternoon, we observed the latter mentioned species attacking a sub-adult male on two separate occasions and seeing it out of its territory. Though extremely small in comparison to the harrier, the wagtail was fearless and pursued the intruder relentlessly until it retreated from the area. A pair of Black-shouldered Kites was also present and as they tend to favour mammalian prey more than small birds, it was a good sign that this area may have a rich abundance of small rodents as well.

Our most obliging airport manager, Rex Topiso had given us permission to remain in the area until after sunset to monitor any harriers that were possibly roosting here and we were not to be disappointed. At around 5.20pm towards the south end of the runway, five harriers, all apparently immature birds began milling over an area of tall grassland approximately 300m from our location. By 5.55pm all of them had gone to ground in this area, with probably others going unseen due to the rapidly fading light. The communal roosting of these birds seemed to suggest for the first time that breeding was not occurring here.

A male Papuan Harrier with a full crop at Mount Hagen Airport   *Rob Simmons*

By the 4th May and the start of our third day of studies, it was now becoming evident that the harriers we had been observing for the past two days were not breeding within the confines of Kagamuga Airport. Birds in a variety of plumages seem to arrive at this location, stay for a few days and take advantage of the prey available to them and presumably move on to their breeding grounds. Only non-breeding birds seemingly remain here permanently. The adult male we observed on the first day was absent the next day, and was replaced the following morning by two different adult males. Similarly, two of the sub-adult males had also left the area. It was therefore unusual to find, what we believed to be, the same two adult females present for a third day. Perhaps, like female Hen Harriers they are the last to arrive at their breeding areas, which were probably in the much higher mid-montane grasslands, a considerable distance from this location.

Our attention then focused on a sub-adult male which was perching on a post in the longish grass a short distance from the main runway. He repeatedly dived down into the long grass on at least four occasions as if he was having difficulty in trying to obtain a prey item which was hidden from our view. Rob went to investigate. We were unable to identify the prey species at the time but scrutiny of our photographic record later that day allowed us to confirm that it was a juvenile Banded Landrail. As Rob had assembled a Bal Chatri trap he decided to keep the young rail as bait to try and catch our old friend, the immature male which was again perching on the metal stand. Our hopes of catching live rodents to use as bait had earlier been dashed due to heavy overnight rain which had flooded the five mammal traps he had set the previous evening. Using live prey is a better option, as the harrier reacts to its movements and then unwittingly gets tangled in the nooses made of strong fishing line This does not injure the bird as it is removed almost immediately for examination. Although the young male did eventually attack the dead Rail a short time later by landing on top of the Bal Chatri, it did not become entangled in the awaiting nooses and flew off when Rob approached it. The prey was then left on top of the metal stand for the bird to eat, and the harrier was seen returning to the same area minutes later.

At 3.30pm we had left the confines of the airport and returned to our hotel as we had arranged to travel into Enga Province, courtesy of the Kumul Lodge minibus, to observe several species of Birds of Paradise and the other beautiful endemics that are to be found in this area, 2,860m asl. You cannot travel to Papua New Guinea, without making the effort to see these amazing birds, and I was hopeful that our visit would be a fruitful and a memorable experience, especially in the shadow of Mount Hagen, which is an awesome 12,579ft high. We travelled to two different locations with our guide Max Poliaka on 5th May to observe the Lesser Bird of Paradise and the King of Saxony Bird of Paradise, at altitudes in montane-forest of 1,600m and 2,500m asl respectively. Other gems included Brown Sicklebill and Ribbon-tailed Astrapia, observed on the garden bird-table at Kumul Lodge. Raptors observed here included Brahminy Kite, Black-mantled and Variable Goshawks, but sadly no Papuan Harriers as suitable grassland habitat was lacking in this area.

The following day, we did not arrive back at Mt Hagen Airport, until mid-afternoon where we were greeted almost immediately at the north end of the runway by what appeared to be a new sub-adult male, with slightly more brown than black markings on his upperwings. Then almost immediately, our attention was drawn once again to the very dark brown and melanistic looking female which was perching low down in a bush only 50m away from our vehicle, which we utilised as a hide, for fear of flushing the bird before we got good views of it. The piercing yellow eyes denoting an adult bird were clearly seen, with the facial disc and head seemingly dark grey, highlighted with a hint of buff when the ruff was occasionally raised around the face. The beak was also grey and did not show a clearly defined and normal yellow cere, which again appeared greyish, with the breast, belly, back and upperwings deep chocolate brown as described previously. The legs and feet were yellow. This was a most attractive bird despite its non-descript plumage markings and probably resembled a dark-morph female Western Marsh Harrier.

The 7th May was our last morning at this location and also for Michael, as he was returning home to Java via Singapore, so this meant we could only spend a couple of hours observing any new harriers that may have arrived overnight. Once again the two birds described above plus the much paler-plumaged female with the patchy brown and buff underwing markings were present. Before leaving the area, Rob and I were of the same opinion that these two females would probably remain here to breed, as they had not moved on during the week and both had been seen associating, albeit briefly, with two sub-adult males. Another raptor we regularly observed here was the Brown Falcon and the 7th was no exception. It patrolled the periphery of the runway at great speed in search of small birds and other quarry.

Just before noon we set off for Goroka in the Eastern Highlands and this journey took almost four hours due to poor road conditions and ended with me sustaining a painful back problem. After a brief visit to the Wildlife Conservation Society offices, we arrived at our accommodation in Goroka Preparatory School, run by Brendan and Pippa Ellis. This would be our home for almost three days prior to John and I returning to the UK. This was probably my last chance of finding breeding Papuan Harriers.

On 8th May our driver Onika from WCS, arrived at 9.00am and immediately drove us to an area of grassland south of Goroka which was convenient to the local golf club. Unfortunately, our watch here was in a low-lying valley and was therefore impeded by tall trees, but we did see at least one Papuan Harrier, probably a female, soaring in tandem with a pair of displaying and possibly Variable Goshawks. An hour later, we found ourselves at 1,500m asl and peering out over an extensive area of grassland adjacent to Kenimora village in the Uggai/Bena District. Our visit immediately attracted the locals, young and old, to join us.

Within minutes of our arrival, an adult female passed low over the grassland directly in front of us and was then joined by a much smaller bird which was almost certainly a male. Both birds began soaring, climbing to a great height after which they were suddenly lost from view amongst the grey rolling clouds. This behaviour is quite common in harriers prior to the breeding season. Then at 2.00pm a beautiful adult male appeared from our left and began foraging close to our observation point which was high up on the roadside, with the valley several hundred metres below us. Over a 15-minute period he made five unsuccessful attempts at catching prey in the long grass. The sixth and final attempt was successful, with a mouse eventually falling victim to his sharp talons. A second adult male was also seen in the vicinity, directly behind the previously described individual. This bird remained high up on the slightly elevated hillside. All our previous harrier sightings were made from the local airports, so our observations that day were highly significant, as these birds were now inhabiting what would probably be their breeding grounds. We remained extremely vigilant in this area for several hours hoping to find that somewhat elusive Papuan Harrier nest, but lady-luck was not shining on us that day.

Male Papuan Harriers   *Rob Simmons*

The 9th May was my last full day in PNG and similarly for John, but Rob still had five days to look for nesting harriers elsewhere. It had been hinted that the Markham Valley in Madang Province, was a potential area for nesting birds. During the afternoon we had to be content with a visit to Goroka Airport, where we were surprisingly allowed to stand on the walkway outside the control tower, to view the 2km airstrip for resident harriers. Similar to the airport at Mt Hagen, the grass was being cut around the periphery of the runway with numerous Black Kites following the tractor and then landing to consume what were thought to be probably frogs and insects disturbed by the operator. There was only one immature male harrier present. It too, took the opportunity to grab a free meal, along with a Black-shouldered Kite, presumably catching any small rodents that suddenly found themselves vulnerable on the bare open ground.

As a treat for our last evening together Brendan and Pippa had arranged for us to be driven to a location in the highlands where we would be cooking and sleeping overnight in a traditional Yabiufa Village straw hut, at 1,930m asl. There was very little grassland available at this altitude, only montane forest, so we knew harriers were unlikely to be observed there. I did, however get a fleeting glimpse of a superb white-phase Variable Goshawk as it dashed across the track in front of our vehicle. The track up to this village was muddy, extremely rough and partially washed away in places creating deep cuts in the sloping ground, but our competent driver David Segeye, negotiated it safely in less than an hour. Little did we know at the time, what the following morning would be like on this track as we travelled back to Goroka!

Our village friends, Kimmy, Oscar, Mike and Yombu, soon got a fire going in a purpose-built shelter and a pre-cooked meal of chicken stew and rice soon filled our empty bellies, washed down with a cup of hot tea. The craic was good around the campfire, as we reminisced about our harrier experiences and the success of the trip so far and speculated on the experiences which might be in store for Rob during the next few days. By 7.30pm heavy rain began to fall, which is apparently common in this area at such a high altitude and by 8.45pm due to its severity, we were forced to retire to our hut. For ten consecutive hours it rained non-stop and had only just ceased when Kimmy and Oscar arrived at around 5.45am to see us safely out of the area. They told us that we would have to walk for several kilometres to meet up with our driver due to the state of the track, as he would be unable to reach the village in such conditions. In places, we were up to our ankles in red clay which clung to our boots like glue, slipping and sliding was inevitable, but this was only the start of a treacherous descent back to our destination. After 3km we eventually met up with David who was waiting for us at another village much lower down the mountain, slowly and at times precariously he got John and I safely back to Goroka almost two hours later, with Rob remaining in the area until later that day. This was a unique Papuan experience for all three of us, which I for one would not forget in a hurry.

When John and I took our leave and bade everyone at Goroka Preparatory School our fond goodbyes and thanks for looking after us, our host Pippa drove us the short distance

Top: Female Papuan Harrier   Bottom: Immature female Papuan Harrier at Mount Hagen   *Rob Simmons*

to the airport to await our flight to Port Moresby. A final scan of the airport, with our binoculars revealed the immature male was still present at 10.00am and prior to us boarding around 10.45am, he was giving us superb views of his foraging skills at the north end of the runway.

On arrival at POM International Airport an hour later, we discovered our onward flight to Singapore had been delayed eight hours and when we were sent to the Airways Hotel restaurant for a complimentary meal, our table on the open balcony conveniently overlooked the 3-4km long runway. To our surprise there were harriers here also, with up to three individuals present, one adult female and two very dark immature birds. They diligently patrolled the extensive area of grassland on either side of the two runways and between the central reservations until sunset, with the fading light eventually beating us. Harriers right up until the end of our journey, was a welcome diversion from the ongoing delays that lay ahead. These resulted in an unscheduled 24-hour visit to Singapore and not being home in time to celebrate Linda's birthday on 11th May!

As Rob, still had five days to search for nesting Papuan Harriers, he did not leave Goroka until mid-day on 12th May, travelling northeast towards the Markham Valley and subsequently to the Ramu Valley, approximately 20km away. The Ramu Valley is to be found two hours west of the City of Lae, on the northeast coast and both these regions are renowned for their extensive grasslands. If harriers were already nesting, then these were the areas in which he was likely to find them. Local knowledge was vital, so his guide for this particular trip was Powa Limu, the head safety and environmental officer for the Ramu Sugar Company.

On 13th May, after several hours of persistent observations in the Markham Valley, Rob found the nest of a Papuan Harrier. It contained two small chicks and an unhatched egg. Having no callipers to measure the egg's vital statistics, Rob returned the following day with the instrument and found the egg had hatched overnight. On 15th May, while cold-searching in the Ramu Valley, another nest was unexpectedly found, it also contained three chicks. Both nests were in reeds, in damp, but not wet ground and were 400-500m asl. There were also several birds displaying at both locations and if Rob had not been pushed for time he reckoned that he would have discovered at least another three nests. The grasslands also contained large numbers of Brown Quail – so rich pickings for breeding Papuan Harriers! These finds were not only historic, but highly significant, as no nest of the Papuan Harrier had been discovered in New Guinea, since 1978, when R Campbell discovered a nest containing two well-grown chicks and an addled egg on 10th May, in grassland at Kagamuga, Mt Hagen (Coates 1985). The location was probably the local airport, where we had commenced our studies of this species on 2nd May.

Other observations made by Rob concluded, that adults have extraordinarily long legs, which is possibly an adaptation to enable them to utilise the long grass for hunting and nesting. This trait was also reflected in the legs of the chicks on the two nests he had visited. During the pre-nuptial displays adults were noted not just lowering their legs but

Top: "Two Spot" – a sub-adult male Papuan Harrier   Bottom: A rare melanistic Papuan Harrier   *Rob Simmons*

fully extending them when coming in to pre-roost. When the sky-dancing displays had finished they would parachute the last 10m in similar fashion into a potential nest site. When in Australia in 2006, the leg-dropping was quite common in the Spotted Harriers I observed, but Rob reckons this trait is recorded with more regularity in the marsh harrier family, as the legs and feet are often dropped in threat displays. This behaviour is also common in Northern and African Marsh Harriers, but not in Black Harriers. Papuan Harriers, however, seem to behave in this manner towards the latter stages of their territorial displays.

Threats to this species are mainly caused by human interests, in the form of the maniacal burning of the grasslands that commences as soon as the dry season arrives. When Rob arrived in this area he literally described the hillsides to me as green, but within the space of three days over 50% of it had been burned. His guide Powa Limu informed Rob, that this is a national pastime in Papua New Guinea and by the end of the season the surrounding hills and much of the lowland valleys would be destroyed by fire. As both are experienced conservationists, their concerns were obviously for the nesting harriers, but whether the nests survived the full term of the breeding season remains doubtful. Leo Legra from WCS agreed to visit both locations to check on the progress of the nests and other nesting harriers. Hopefully the outcome was successful.

Probably, the most important thing to emerge from this whole trip was the collection of two vital blood samples, obtained from a single chick from each nest, to establish from genetic investigation if this is a true species of marsh harrier. But, this is Rob's special project and I dare not intrude in his scientific studies, so I, like everyone else, will have to wait for his conclusion which I expect will appear in print in due course.

Another interesting snippet of information worth mentioning is – that these are very big birds and may be comparable and possibly heavier and larger than the neighbouring Australian Swamp Harrier, which is presently regarded as the world's heaviest harrier! But until live adult specimens are caught, accurately weighed and egg measurements taken we can only speculate at present. A fascinating story for another day.

Prior to visiting PNG, I had reservations concerning my (our) safety, after reading disturbing information on the internet and a certain tourist guide book. Well to allay everyone's fears, and contrary to this, I had a superb time and was treated reverently wherever I went. This goes to show that you should not believe everything you read! It is a beautiful island, and much of it is still to be explored. I met a lot of new friends and observed some unique scenery and wildlife, which must be visited at least once and in my case hopefully twice, in one's lifetime!

Melanistic Papuan Harrier in the Highlands of New Guinea   *Rob Simmons*

13

FINALE AND AFTERTHOUGHTS

Which is my favourite harrier species?

I was asked this question many times prior to my visit to Papua New Guinea and on my return, after completing my quest to observe all 16 species of harrier.

The answer has to be the Hen Harrier, as I have spent 22 consecutive seasons with this species in my native Northern Ireland. If I did not have this elegant and attractive raptor to study here, I doubt if I would even bother to go out bird-watching. But, there are three other species of harrier, which I personally regard as equally attractive. In two of these species, the males and females sport similar plumages. These are the Black Harrier from South Africa and the Spotted Harrier from Australia. The third species is the male Pied Harrier from eastern Asia which is another stunning example.

When I first observed these three harriers in 2002, 2006 and 2003 respectively, they were the ones, along with the Hen Harrier, which made the hair stand up on the back of my neck. The larger species of this family of raptors are awesome birds in their own right. The largest, the Long-winged Harrier from South America, is a spectacular bird in flight. The heaviest, the Swamp Harrier from Australia, and the slightly lighter Madagascar Marsh Harrier, are both a match for most large predators, in comparison to the smallest and lightest of all the harriers, the butterfly-like and dainty little Montagu's. As a consequence of studying in the wild, all members of the *Circus* family, I truly regard each species as equal, with of course the odd exception, plumage-wise, which sets these birds poles apart from most other raptors.

Greatest harrier moments

When I embarked on my study of the world's harrier species over two decades ago, it was probably inevitable that I would have many memorable and historic experiences with this unique family of raptors. These 'never to be forgotten' and magical moments are too numerous to mention individually and so I specifically select five, which in my opinion stand out from the rest.

Observing my first male and female Hen Harriers, in the Antrim Plateau during the spring of 1986, with aerobatic sky-dancing displays and spectacular food-passes performed before my very eyes. I was immediately hooked!

Discovering the first Hen Harrier tree nest, on 14th July 1991, at a mature conifer plantation in South Antrim. This find ranks high on my list, as it was not only an amazing and evolutionary feat, but also an ornithological first for this species!

Being part of the team of five observers, that counted the largest ever mixed roost of harriers in the world (over 3,000 birds), at Velavadar Blackbuck National Park, Gujarat District, NW India, on 6th December 1997. This was an incredible sight, which will probably never be equalled again and ranks as one of my greatest ever harrier moments!

Achieving my long-standing ambition in October 2006, to visit Australia to specifically study the Spotted Harrier, which habitually nests in trees and is the world's original tree-nesting harrier. What a bird!

Fulfilling my 'Harrier Journeys' in May 2007, with the study of the little known Papuan Harrier on the alluring and mysterious island of New Guinea, which gave me a feeling of complete satisfaction and sheer relief that the final leg of my quest had ended successfully.

Dreams do come true if you persist!

Conservation concerns worldwide

As Rob Simmons and David Hollands rightly point out in their respective *Foreword* and *Preface* at the beginning of this book, almost all of the world's harrier species could be regarded as endangered and highly vulnerable, with several prime candidates for global extinction. Those found solely on islands appear to be the most vulnerable namely, the Madagascar, Reunion and Papuan Harriers. The endemic Black Harrier from South Africa and the subspecies *Harterti,* which has a restricted range in North Africa, may also be globally threatened. Working on the basis of the 17 species individually mentioned in this book, it is frightening to think that 29% of this family of raptor is in danger of becoming extinct within a generation and possibly sooner, if steps are not taken now to conserve them.

It is also known that other members of the genus, such as Pallid and Montagu's are decreasing and becoming localised within their European range as is the more common Hen Harrier, particularly in parts of the UK uplands. The African Marsh is also decreasing on the continent of Africa. Little is really known of several other species, like the Long-winged and Cinereous Harriers from South America and if the truth be known, the Pied and Eastern Marsh, from Asia. What is known however, is that the Western or European Marsh Harrier is increasing rapidly in parts of Europe and the UK and will hopefully re-colonise my native Northern Ireland and the Republic of Ireland in the near future. Numbers of Northern Harriers in the USA remain strong, as do Swamp Harriers in Australia and New Zealand, but Spotted Harriers are sparsely distributed throughout Australia in comparison, and therefore their status is difficult to assess, nevertheless, the population is currently regarded as stable.

The above assessment shows that all harrier species appear to be unremittingly vulnerable to either loss of natural habitat, shooting on migration, persecution at their breeding grounds, or in the case of Montagu's, a casualty of harvesting operations, pesticide poisoning in their wintering grounds, and more recently drought. Irreversible changes and losses of their natural habitat has forced them to nest and winter in areas which have not been given protective designation. This too renders them vulnerable to man and is undoubtedly a significant reason for their overall decline in their countries of origin. As a result, several species still appear to be abundant, but now only in localised areas. Further changes to these places, which have been occupied by harriers as a last resort, could be catastrophic to the overall survival of the majority of the world's harrier species.

Most recently, reviews of several species, undertaken within their respective countries by reputable conservation organisations, universities and concerned ornithologists, many of them harrier experts, have now recommended immediate protection to not only the species itself, but also to its vital habitat, for without the latter their demise would surely be imminent. For example, in the Southern Cape region of South Africa, ongoing studies of Black and African Marsh remain a priority for the University of Cape Town. On Madagascar too, a recent status review of the Madagascar Marsh Harrier has prompted the Peregrine Fund, to take immediate action to save this species from extinction. French ornithologists on nearby Reunion Island, are also worried about the long term future of the appropriately named Reunion Harrier. After our recent return from Papua New Guinea, Rob Simmons and I had similar worries concerning the future of the Papuan Harrier, and these have been relayed to the locally based Wildlife Conservation Society and other relevant organisations within New Guinea. Creating awareness through education has been implemented by several of these reputable bodies in a last ditch effort to conserve these endemics. Hopefully their efforts have not come too late.

I do not wish to paint a picture of doom and gloom for harrier species worldwide, but as David Hollands correctly says in the *Preface* of this book – harriers have never enjoyed the celebrity status of eagles or falcons and apart from a few species, very little is really known about them. I can vouch for every word of David's perceptive statement, regarding my experiences with the Hen Harrier in Northern Ireland. Hopefully, long term conservation measures and stringent laws to fully protect harriers worldwide will be implemented soon, particularly by responsible birding organisations, working hand-in-hand with local governments, within countries where their resident species of *Circus,* is at risk of extinction. I hope Rob Simmons' sobering remarks earlier in this book, 'that in 25 years or perhaps less, harriers may only survive in protected Reserves', never comes true, as these denizens of wide open spaces and wetlands deserve much more from us, don't they? Protecting the environment and its wildlife is an obligation, not a choice!

Conclusion

Before bringing final closure to this book, I could not resist mentioning briefly my visits to Dumfries and Galloway in search of nesting Hen Harriers and to Leighton Moss in Lancashire, to observe breeding Western Marsh Harriers. Both these areas were visited between 11th and 15th June 2007. At the latter location a polygynous male successfully held court with two willing females. At the first nest, two fledglings were very active above the reedbed, but the unknown number of chicks had not yet fledged from the second site. During my sojourn in southern Scotland, I purposely took time out from my birding activities to visit the grave of harrier colleague Donald Watson, who died in November 2005 and is buried in St John's Town of Dalry. Sadly, the Hen Harriers in this area did not appear to have had a successful season, as I observed two grey males still sky-dancing on 11th and 15th June respectively, with not a female to be seen!

On return home, I visited four known Hen Harrier locations in South Antrim on 18th June, but activity was minimal in all these areas, which was probably due to the continuous heavy rain that badly affected Northern Ireland, for five full days between 12th and 16th June and thereafter. I fear therefore, that several ground and tree nests may have been completely lost, as the latter tend to collapse and chicks inevitably die on the forest floor. When climate change finally gives the UK and Ireland, much drier and consistently warmer weather conditions, then the success rate of this species is likely to improve substantially, but for the present it is a case of wait and see. It would also appear that there are dark and sinister forces operating in the northern climes of the Antrim Plateau, to ensure that Hen Harrier numbers do not increase. Declines in this area alone have been significant in recent years, despite the lack of Red Grouse and the unjustified propaganda blaming this species for its demise issued of course, by the so-called shooting fraternity.

In 2006 there were only 14 pairs of Hen Harriers in the Antrim Hills, many of which failed to breed or produced small and insignificant broods which were unlikely to sustain a viable breeding population in this area. In 2007 the number declined to 12 pairs and by 29th July only three nests were known to have been successful, producing just five young. If I was allowed to ask one specific question, it would have to be to the RSPB. Could you please explain – where have over 50% of our Hen Harriers gone in three years? My honest answer to that important question is – that the numbers they came up with during the 2004 breeding survey, never really existed did they, especially in County Antrim? Despite all these problems I have to remain optimistic, that this species will ultimately be successful in the near future; otherwise we may lose this beautiful raptor, again!

Finally

My final tribute to these wonderful birds was put to me in the form of a question by my long-suffering wife Linda. Now that you have been all round the world and studied all these species of harrier, what do you intend to do now? My instant reply was – go round the world and see them all again! To which she replied – surely not! I also find it stimulating that there is still a lot to be discovered in nature. Who knows, there may be another unknown species or subspecies of harrier, probably on a remote island somewhere in the world that scientists and taxonomists have not yet studied. It has been rumoured that there is a harrier on the island of Fiji, in the South Pacific, that has not yet been studied, so I await confirmation as to its true identity. My passport is getting dusty in the back of my chest of drawers – could this be my next destination?

Having clocked up over 224,000 air miles, through domestic, internal flights within destinations and long haul travel, the equivalent of having circumnavigated the world nine times and not including motorised travel within countries visited, I thank God, for allowing me to complete my 'Harrier Journeys', safely and successfully!

Books related to harriers

Clarke, R. 1990. *Harriers of the British Isles.* Shire, Princes Risborough.

Clarke, R. 1995. *The Marsh Harrier.* Hamlyn, London.

Clarke, R. 1996. *Montagu's Harrier.* Arlequin Press, Chelmsford.

Hamerstrom, F. 1986. *Harrier, Hawk of the Marshes.* Smithsonian Press, Washington.

Simmons, R. E. 2000. *Harriers of the World.* Oxford University Press, Oxford.

Verma, A. 2007. *Harriers in India: A Field Guide.* Wildlife Institute of India, Dehradu.

Watson, D. 1977. *The Hen Harrier.* Poyser, Berkhamsted.

Weis, H. 1923. *Life of the Harrier in Denmark.* Wheldon & Wesley, London.

General Bibliography

Baker-Gabb, D. J. 1984. *The Evolution of Tree-nesting and the Origin of the Spotted Harrier.* Corella 8: 67-69.

Balfour, E. & Cadbury, C. J. 1979. *Polygyny, spacing and sex ratio among Hen Harriers in Orkney, Scotland.* Ornis Scandinavica 10: 133-141.

Barua, M. & Sharma, P. 1999. *Birds of Kaziranga National Park, India.* Forktail 15: 47-60.

Bo, M. S. Cicchino, S. M. & Martinez, M. M. 1996.*Diet of Long-winged Harrier in South-Eastern Buenos Aires Province, Argentina.* Jour. Raptor Research 30: 237-239.

Bourne, W. R. P. 1992. *What happens when Hen Harriers nest above ground?* Irish Birds 4: 564-565.

Bretagnolle, V. Ghestemme, T. Thiollay, J-M. & Attie, C. 2000. *Distribution, population size and habitat use of the Reunion Marsh Harrier.* Forktail 34: 8-17.

Bretagnolle, V. Thiollay, J-M. & Attie, C. 2000. *Status of Reunion* Marsh *Harrier on Reunion Island.* Chancellor, R. & Meyburg, B.-U. eds. 2000. Raptors at Risk. WWGBP, Hancock House. WA.

Chadwick, P. 1997. *Breeding by Black Harriers in the West Coast National Park, South Africa.* Jour. African Raptor Biology 12: 14-19.

Clark, W. S. 1997. *A dark morph for the African Marsh Harrier.* Jour. African Raptor Biology 12: 27-29.

Clarke, R. Bourgonje, A. & Castelijns, H. 1993. *Food niches of sympatric Marsh Harriers and Hen Harriers on the Dutch coast in winter.* Ibis135: 424-431.

Clarke, R. Prakash, V. Clark, W.S. Ramesh, N. Scott, D. 1998. *World record count of roosting harriers in Blackbuck National Park, Velavadar, Gujarat, north-west India.* Forktail 14: 70-71.

Coates, B. J. 1985. *The Birds of Papua New Guinea. Vol.1.* Dove Publications, Queensland.

Cormier, J-P. & Baillon, F. 1991. *Concentration de Busards cendres Circus pygargus (L) dans la region de M'Bour (Senegal) Durant L'hiver 1988-89:* Utilisation du milieu et regime alimentaire. *Alauda 59: 163-168.*

Crooke, C. H. 1996. *First record of breeding Marsh Harrier for the Highland Region.* Highland Bird Report: 51.

Cupper, J & Cupper, L. 1981. *Hawks in Focus.* Jaclin Enterprises, Mildura.

Curtis, O. Jenkins, A. & Simmons, R. 2001. *The Black Harrier – Work in progress.* Africa Birds & Birding 6, October/November Issue: 30-36.

D'Arcy, G. 1988. *The Animals of Ireland.* Appletree Press, Belfast.

Deane, C.D. 1954. *Handbook of the Birds of Northern Ireland.* Belfast Museum Publication.

De La Pena, M. R. & Rumboll, M. 1998. *Birds of Southern South America and Antarctica.* Harper Collins. London.

Fairclough, K. 1995. *The Orkney Pallid Harrier.* Orkney Bird Report: 67-68.

Fefelov, I. V. 2001. *Comparative breeding ecology and hybridization of Eastern and Western Marsh Harriers, in the Baikal region of Eastern Siberia.* Ibis 143 (4): 587-592.

Ferguson-Lees, J. & Christie, D.A. 2001. *Raptors of the* World. Christopher Helm. London.

Goodman, S. M. & Benstead, J. P. 2003. *The Natural History of Madagascar.* University of Chicago Press. Chicago.

Hedley, L. A. 1985. *Another example of Tree-Nesting (Swamp Harriers).* Notornis 32: 22.

Hesketh, B. & Murphy, B. 2004. *Female harrier the belle of the moor.* Bird Watching Magazine: June Issue, 2004.

Hollands, D. 2003. *Eagles Hawks and Falcons of Australia.* Bloomings Books. Melbourne.

Hugo, S. 2006. *Marsh Harriers soar to new heights.* Bird Watching, July Issue: 22-23.

Hutchinson, C. D. 1989. *Birds in Ireland.* Poyser, Calton.

Kemp, A. & Dean, R. 1988. *Diet of African Marsh Harriers from pellets.* Gabar 3: 54-55.

Kemp, A. & Kemp, M. 2001. *Birds of Prey of Africa and its Islands.* Struik Publishers. New Holland. Cape Town.

Langrand, O. 1990. *Guide to the Birds of Madagascar.* Yale University Press. New Haven & London.

McCurdy, K. M. Orr, S. J. & Hodgkins, T. M. 1995. *A Large Northern Harrier Roost at Fort Sill, Oklahoma.* Bulletin of the Oklahoma Ornithological Society 28: 25-27.

McMillan, R. L. 2007. *Hen Harrier Monitoring Study: Skye 2007.* skye-birds.com, Elgol, Isle of Skye.

Mellon, C. Allen. D. & Scott, D. 2005. *Proposed Special Protection Areas for Hen Harriers in Northern Ireland.* Environment & Heritage Service Report: 1-41.

Messenger, B. 1990. *Tree-nesting Harriers (Swamp Harriers).* Notornis 27: 172.

Morris, P. & Hawkins, F. 1998. *Birds of Madagascar.* Pica Press. East Sussex.

Naoroji, R. 2006. *Birds of Prey of the Indian Subcontinent.* Christopher Helm. London.

Paz, U. 1987. *The Birds of Israel.* Christopher Helm. London.

Picozzi, N. 1984. *Breeding biology of polygynous Hen Harriers in Orkney.* Ornis Scandinavica 15: 1-10.

Rene de Roland, L-A. Rabearivony, J. Randriamanga, I. & Thorstrom, R. 2004. *Nesting biology and diet of the Madagascar Harrier in Ambohitantely Special Reserve, Madagascar.* Journal Raptor Research 38: 256-262.

Scott, D. Clarke, R. & Shawyer, C. R. 1991. *Hen Harriers breeding in a tree nest.* Irish Birds 4: 413-417.

Scott, D. Clarke, R. & McHaffie, P. 1992. *Hen Harriers successfully breeding in a tree nest of their own construction.* Irish Birds4: 566-570.

Scott, D. 1993 *Hen Harriers tree nest again in 1993.* Northern Ireland Bird Report: 69-71. N.I.B.A.

Scott, D. Clarke, R. & Shawyer, C. R. 1993/ 94. *Tree-nesting Hen Harriers –Evolution in the making?. The Raptor21: 53-56.*

Scott, D. 1994. *Summer 1994 – Observations of Marsh and Montagu's Harriers at the S'Albufera, Mallorca.* The Harrier, Newsletter of the Northern Ireland Ornithologists Club. Spring 1995: 5-6.

Scott, D. & Clarke, R. 1995. *First records of tree roosting by Hen Harriers in Northern Ireland.* Northern Ireland Bird Report : 84-89. N.I.B.A.

Scott, D. & Clarke, R.1995. *The Marsh Harrier in Northern Ireland.* Northern Ireland Bird Report: 82-87. N.I.B.A.

Scott, D. 1996. *Records of Albinism in Irelands Hen Harrier Population.* Northern Ireland Bird Report: 101-106. N.I.B.A.

Scott, D. 1998. *The Ecology of a Hen Harrier Winter Roost.* Northern Ireland Bird Report: 105-116. N.I.B.A.

Scott, D. & Scott, L. 1998. *Melanistic Male Hen Harrier in south-west Scotland.* Unpublished Report.

Scott, D. 1999. *Influx of Short-eared Owls and Hen Harriers into Northern Ireland during 1999.* Northern Ireland Bird Report: 120-127. N.I.B.A.

Scott, D. 1999. *A comparison of the status of the Marsh Harrier in Northern Ireland and the Isle of Man.* Peregrine 7: 475-481. Manx Ornithological Society.

Scott, D. 2000. *Marking a decade of tree nesting by Hen Harriers in Northern Ireland, 1991-2000.* Irish Birds 6: 586-589.

Scott, D. 2002. *How many species of Harrier are there? – and my quest to find them all.* Northern Ireland bird Report: 126-129. N.I.B.A.

Scott, D. & McHaffie, P. 2003. *What impact do Buzzard and Goshawk have on other raptors in coniferous forest? – preliminary findings.* Irish Birds 7. 267-269.

Scott, D. 2004. *Four in a row for tree nesting Hen Harriers in 2002.* Irish Birds 7: 440.

Scott, D. 2005. *The diet of Hen Harriers in Northern Ireland.* Irish Birds 7: 597-599.

Scott, D. 2006.*Roosting Northern Harriers at Merritt Island NWR, Florida.* Report to Management & Habi-Chat Newsletter Briefing, Summer 2006: 5.

Scott, D. & Hipkiss, T. 2006. *Tree-nesting behaviour by a polyandrous female Hen Harrier.* Irish Birds 8: 139-141.

Scott, D. & Clarke, R. 2007. *Comparing the success of Hen Harrier tree nests and ground nests in the Antrim Hills, 1990-2006.* Irish Birds 8: 315-318

Simmons, R. E. Curtis, O. & Jenkins, A. R. 2000. *Black Harrier conservation and ecology: preliminary findings 2000.* Jour. Afr. Raptor Biol. 13: 33-38.

Simmons, R. E. 2000. *Harriers – Skydancing through time.* Africa Birds & Birding 5, October/November Issue: 36-43.

Skerrett, A. & Roest, L. 2002. *Pallid Harrier: the first record for Seychelles.* African Bird Club 10-2: August 2003, 126-127.

Skinner, J. F. 1979. Puzzling Behaviour of Harriers. Notornis 26: 119.

Steyn, P. 1982. *Birds of Prey of Southern Africa.* David Philip, Cape Town & Johannesburg.

Thomas, M. 2006. *Montagu's prosper in 2006.* Birdwatch Magazine: October Issue, 22.

Thorpe, J. P. & Kelly, G. M. 1995. *A Black Hen Harrier at Ballaugh Curragh.* Peregrine 7: 289-290.

Underhill-Day, J. 1998. *Breeding Marsh Harriers in the United Kingdom, 1983-95.* British Birds 91: 210-218.

Watson, D. 1991. *Hen Harriers breeding in a tree nest: further comments.* Irish Birds 4: 418-420.

Zahavi, A. 1957. *The breeding birds of the Huleh Swamp and Lake (Northern Israel).* Ibis 99: 600-607.

Birds

African Fish-eagle *Haliaeetus vocifer*
African Marsh Harrier *Circus ranivorus*
American Coot *Fulica americana*
American Kestrel *Falco sparverius*
Aplomado Falcon *Falco femoralis*
Australian Hobby *Falco longipennis*
Australian Magpie *Gymnorhina tibicen*
Australian (Richard's) Pipit *Anthus novaeseelandiae*
Bald Eagle *Haliaeetus leucocephalus*
Banded Landrail *Rallus philippensis*
Barn Owl *Tyto alba*
Black Harrier *Circus maurus*
Black-headed Gull *Larus ridibundus*
Black Kite *Milvus migrans*
Black-mantled Goshawk *Accipiter melanochlamys*
Black-shouldered Kite *Elanus caeruleus*
Black Vulture *Coragyps atratus*
Black-winged Stilt *Himantopus himantopus*
Blue-eyed Cormorant *Phalacrocorax atriceps*
Blue-winged Teal *Anas discors*
Boat-tailed Grackle *Quiscalus major*
Booted Eagle *Hieraaetus pennatus*
Brahminy Kite *Haliastur indus*
Brown Falcon *Falco berigora*
Brown Goshawk *Accipiter fasciatus*
Brown Quail *Coturnix ypsilophora*
Brown Sicklebill *Epimachus mayeri*
Cape Francolin *Francolinus capensis*
Chimango Caracara *Milvago chimango*
Cinereous Harrier *Circus cinereus*
Common Buzzard *Buteo buteo*
Common Kestrel *Falco tinnunculus*
Common Teal *Anas crecca*
Coot *Fulica atra*
Crested Caracara *Polyborus plancus*
Crested Serpent Eagle *Spilornis cheela*
Curlew *Numenius arquata*
Eastern Marsh Harrier *Circus spilonotus*
Elegant Crested Tinamou *Eudromia elegans*

Eleonora's Falcon *Falco eleonorae*
Elephant Bird *Aepyornis maximus*
Eurasian Eagle Owl *Bubo bubo*
Galah *Cacatua roseicailla*
Golden Eagle *Aquila chrysaetos*
Goshawk *Accipiter gentilis*
Great Grey Owl *Strix nebulosa*
Great Horned Owl *Bubo virginianus*
Great Indian Bustard *Choriotis nigriceps*
Greater Rhea *Rhea americana*
Green-winged Teal *Anas carolinensis*
Greylag Goose *Anser anser*
Hen Harrier *Circus cyaneus*
Herring Gull *Larus argentatus*
Hooded Crow *Corvus cornix*
Hooded Mannikin *Lonchura spectabilis*
King of Saxony Bird of Paradise *Pteridophora alberti*
Lesser Bird of Paradise *Paradisea minor*
Lesser Rhea *Pterocnemia pennata*
Little Button-quail *Turnix velox*
Long-winged Harrier *Circus buffoni*
Madagascar Buzzard *Buteo brachypterus*
Madagascar Harrier-Hawk *Polyboroides radiatus*
Madagascar Marsh Harrier *Circus macrosceles*
Madagascar Pochard *Aythya Innotata*
Magellanic Penguin *Spheniscus magellanicus*
Mallard *Anas platyrhynchos*
Marsh Owl *Asio capensis*
Meadow Pipit *Anthus pratensis*
Merlin *Falco columbarius*
Montagu's Harrier *Circus pygargus*
Moorhen *Gallinula chloropus*
Nankeen Kestrel *Falco cenchroides*
North-African race of Western Marsh Harrier *Circus aeruginosus harterti*
Northern Giant Petrel *Macronectes halli*
Northern Harrier *Circus hudsonius*
Osprey *Pandion haliaetus*
Pallas's Fishing Eagle *Haliaeetus leucoryphus*
Pallid Harrier *Circus macrourus*
Papuan Harrier *Circus spilothorax*
Peregrine Falcon *Falco peregrinus*

Pied Chat *Saxicola caprata*
Pied Crow *Corvus albus*
Pied Harrier *Circus melanoleucos*
Powerful Owl *Ninox strenua*
Raven *Corvus corax*
Red-backed Hawk *Buteo polyosoma*
Red-breasted Sparrowhawk *Accipiter rufiventris*
Red Goshawk *Erythrotriorchis radiatus*
Red Grouse *Lagopus lagopus hibernicus*
Red-winged Blackbird *Agelaius phoeniceus*
Reunion Harrier *Circus maillardi*
Ribbon-tailed Astrapia *Astrapia mayeri*
Richards Pipit *Anthus novaeseelandiae*
Roadside hawk *Buteo magnirostris*
Savannah Hawk *Heterospizias meridionalis*
Secretarybird *Sagittarius serpentarius*
Short-eared Owl *Asio flammeus*
Skylark *Alauda arvensis*
Snail Kite *Rostrhamus sociabilis*
Snipe *Gallinago gallinago*
Sooty Owl *Tyto tenebricosa*
Southern Boobook Owl *Ninox novaeseelandiae*
Sparrowhawk *Accipiter nisus*
Spotted Harrier *Circus assimilis*
Starling *Sturnus vulgaris*
Swamp Harrier *Circus approximans*
Tawny Grassbird *Megalurus timorensis*
Turkey Vulture *Cathartes aura*
Variable Goshawk *Accipiter novaehollandiae*
Wedge-tailed Eagle *Aquila audax*
Western Marsh Harrier *Circus aeruginosus*
Whistling Kite *Haliastur sphenurus*
White-faced Ibis *Plegadis chihi*
White-fronted Goose *Anser albifrons*
White Ibis *Eudocimus albus*
Willie Wagtail *Rhipidura albolimbata*

Mammals

Asian Short-clawed Otter *Amblonyx cinereous*
Asiatic Wild Buffalo *Bubalus bubalis*
Badger *Meles meles*
Bengal Tiger *Panthera tigris tigris*
Blackbuck *Antilope cervicapra*
Bobcat *Felis rufus*
Brown Lemur *Eulemur fuluus*
Brown Rat *Rattus norvegicus*
Commersons Dolphin *Cephalorhynchus commersonii*
Common Mongoose *Herpestes edwardsi*
Field Vole *Microtus agrestis*
Grey Fox *Urocyon cinereoargenteus*
Guanaco *Lama guanicoe*
Indian Elephant *Elephas maximus*
Indian Fox *Vulpes bengalensis*
Indian Rhinoceros *Rhinoceros unicornis*
Indian Wolf *Canis lupus pallipes*
Irish Hare *Lepus timidus hibernicus*
Irish Stoat *Mustela erminea hibernica*
Jackal *Canis aureus*
Jungle Cat *Felis chaus*
Lemming *Lemmus lemmus*
Leopard *Panthera pardus*
Lesser Bandicoot Rat *Bandicota bengalensis*
Rabbit *Oryctolagus cuniculus*
Red Fox *Vulpes vulpes*
Skunk *Mephitis mephitis*
Striped Mouse *Rhabdomys pumilio*
Vlei Rat *Otomys irroratus*

Reptiles and Amphibians

Common Frog *Rana temporaria*
Frog *Rana spp*
Mainland Tiger Snake *Notechis scutatu*
Viviparous Lizard *Lacerta vivepara*